Politics, Governance and Technology

NEW HORIZONS IN PUBLIC POLICY

General Editor: Wayne Parsons
Professor of Public Policy, Queen Mary and Westfield College, University of London, UK

This series aims to explore the major issues facing academics and practitioners working in the field of public policy at the dawn of a new millennium. It seeks to reflect on where public policy has been, in both theoretical and practical terms, and to prompt debate on where it is going. The series emphasises the need to understand public policy in the context of international developments and global change. New Horizons in Public Policy publishes the latest research on the study of the policy making process and public management, and presents original and critical thinking on the policy issues and problems facing modern and post-modern societies.

Titles in the series include:

Innovations in Public Management
Perspectives from East and West Europe
Edited by Tony Verheijen and David Coombes

Public Policy Instruments
Evaluating the Tools of Public Administration
Edited by B. Guy Peters and Frans K.M. van Nispen

Beyond the New Public Management
Changing Ideas and Practices in Governance
Edited by Martin Minogue, Charles Polidano and David Hulme

Economic Decentralization and Public Management Reform
Edited by Maureen Mackintosh and Rathin Roy

Public Policy in the New Europe
Eurogovernance in Theory and Practice
Edited by Fergus Carr and Andrew Massey

Politics, Governance and Technology
A Postmodern Narrative on the Virtual State
P.H.A. Frissen

Public Policy and Political Institutions
The Role of Culture in Traffic Policy
Frank Hendriks

Politics, Governance and Technology

A Postmodern Narrative on the Virtual State

P.H.A. Frissen

Professor of Public Administration, Tilburg University, The Netherlands

Translation: Dr Chris Emery, University of Teesside, UK

NEW HORIZONS IN PUBLIC POLICY

Edward Elgar

Cheltenham, UK • Northampton, MA, USA

© Paul Frissen, 1999.

All rights reserved. No part of this publication may be reproduced, stored in a retrieval system or transmitted in any form or by any means, electronic, mechanical or photocopying, recording, or otherwise without the prior permission of the publisher.

Published by
Edward Elgar Publishing Limited
Glensanda House
Montpellier Parade
Cheltenham
Glos GL50 1UA
UK

Edward Elgar Publishing, Inc.
136 West Street
Suite 202
Northampton
Massachusetts 01060
USA

A catalogue record for this book
is available from the British Library

Library of Congress Cataloguing in Publication Data

Frissen, P.H.A. (Paul H.A.) 1955–
 [Virtuele staat, politiek, bestuur, technologie. English]
 Politics, governance, and technology: a postmodern narrative on the virtual state / P.H.A. Frissen: translation, Chris Emery.
 (New horizons in public policy series)
 Includes bibliographical references.
  1. Political science. 2. Public administration. 3. Information society—Political aspects. 4. Information technology—Political aspects. 5. Postmodernism—Political aspects. I. Title. II. Series: New horizons in public policy.
 JA80.F7413   1999
 320—dc21   99–38281
                                                                                    CIP

Printed and bound in Great Britain by Biddles Ltd, www.Biddles.co.uk

ISBN 1 85898 877 2

To Jef Frissen (1922–1992)

Contents

List of Tables		x
People, Places, Activities: an Acknowledgement		xi
1	**Prologue**	1
1.1	Politics, Administration, Technology: Connections	1
1.2	Postmodernisation and Postmodernism	4
1.3	The Narrative	6
2	**Public Administration**	9
2.1	Change, Experiment, Innovation: an Impression	9
2.2	The New Governance: Characteristics	11
2.3	The Antipode: Classical Governance	13
2.4	The Tradition: Sovereignty, Subsidiarity, Corporatism	14
2.5	A Case: Independent Agencies	16
2.6	Case 2	20
2.7	Amendments to Bureaucracy	23
2.8	Differentiating the Bureaucracy	28
2.9	Reversing Bureaucracy	30
3	**Technology**	36
3.1	Informatisation: an Outline	36
3.2	Informatisation: Developments	38
3.3	Informatisation: Characteristics	40
3.4	The Technological Debate: Positions	42
3.5	Advanced Constructivism or Determinism?	44
3.6	Networks, Connectivity and Virtual Reality	46
3.7	Organisational Meaning	50
3.8	Cultural Meaning	56
3.9	Digital Ambiguities I	61

4	**Public Administration and Technology**	64
4.1	Informatised Bureaucracy: an Impression	64
4.2	Characteristics of Informatisation	70
4.3	Bureaucratisation and Technocratisation: the Perspective of Modernisation	73
4.4	Informatisation, Change, Experiment and Innovation	78
4.5	Transformations	83
4.6	Case 3	88
4.7	The Primacy of Politics	91
4.8	Digital Ambiguities 2	92
5	**Politics**	95
5.1	Crisis and Continuity: Introduction	95
5.2	Debate and Self-Reflection	96
5.3	Restored Values	101
5.4	Bureaucracy and Technocracy: the New Politics	110
5.5	The Empty Place of Power I	117
5.6	Political Strategies	121
5.7	And yet ... the Great Silence	126
6	**The Social Decor: Modernisation and Postmodernisation**	128
6.1	Economic Transformations	129
6.2	Organisational Changes	132
6.3	Cultural Patterns	136
6.4	Images	141
6.5	Advanced Modernisation	144
6.6	Nuances and Discontinuities	151
7	**Theoretical Intermezzo**	160
7.1	Postmodernisation, Postmodernity, Postmodernism	161
7.2	We Postmodern Bourgeois Liberals	164
7.3	Yuppies, Ephemera and Kitsch	169
7.4	Hyperdifferentiation and Fragmentation	175
7.5	Centre and Linearity	180
7.6	Ambiguities	187

8	**Administration and Politics in Postmodern Cyberspace**	190
8.1	Technoculture	191
8.2	Virtualisation	195
8.3	Governance: Intelligence and Aesthetics	202
8.4	Politics: Primacy or the End?	207
8.5	Politics, Governance and Technology: Fragmentation and Connections	213
9	**Fragmenting and Connecting Governance**	219
9.1	Fragmentation and Contingency	220
9.2	Autonomy and Horizontalisation	224
9.3	Once again: Infrastructure	229
9.4	Forms and Styles	235
9.5	Postmodern Administrative Theory	238
10	**Politics without Properties**	244
10.1	Political Theory: a Modern Project	245
10.2	Fragmentation, Contingency, Indifference	249
10.3	The Apolitical Public Domain	255
10.4	Postmodern Politics	259
10.5	Political Postmodernism	264
10.6	Digital Ambiguities 4: the Virtual State	268
Bibliography		272
Index		292

Tables

2.1	Reorganisation of the Civil Service	25
7.1	Modernism and Postmodernism	174

People, Places, Activities: an Acknowledgement

Any narrative is a product of countless connections with different people, places and activities. People contribute through their inspiration, support, disagreement and the pleasure which they give. Places provide the context within which a narrative like this gradually takes shape. Activities are the source of experience, knowledge and insight which feed the narrative's content. But although in this narrative the subject is 'de-centred', I alone am responsible for any shortcomings which the reader may encounter. My heartfelt thanks go to the following people, places and activities.

People

The members of the Tilburg/Rotterdam research programme 'Informatisation in Public Administration' who constantly demonstrate the importance of passion and pleasure.

Pieter Tops who, with the effortlessness of true friendship, enabled me to conceive of and produce this narrative.

Ig Snellen, Danièle Bourcier, Wim van de Donk, Klaus Grimmer, Jeroen van den Hoven, Klaus Lenk, Odette Meyer and Arre Zuurmond: companions in the often turbulent microcosm of *NIAS*.

Marijke Nobel, Vivian Carter and Ann Musters who converted my handwritten manuscript into electronic form.

Saskia Lindner, Arthur van Baalen, Petra Mettau and Bram Foederer who assembled the literature and performed numerous other services.

Chris Emery who succeeded in translating typically Dutch writing into English.

Marijke van der Putten, Renée Frissen and Rutger Frissen who are a reality which is a grand narrative in itself.

Places

Tilburg University whose virtual nature, southern character and disordered freedom create a perfect environment for narrators.

The Netherlands Institute for Advanced Study in the Humanities and Social Sciences (NIAS) at Wassenaar which enabled me to spend a year (1993–94) thinking and reading. This narrative is an expression of my gratitude.

Public Administration in the Netherlands, postmodern or not, is a fascinating phenomenon about which I speak and write with passion and pleasure.

Cyberspace which has turned a dyed-in-the-wool 'arts' man into a lover of the virtual reality of information and communication technology.

Activities

Research, teaching, consultancy, management, lectures, conferences, public debate are the activities in which this narrative has been tested, refined, sharpened, rejected, tried out and nurtured.

Writing is that miraculous process – almost an *'écriture automatique'* – by which this narrative finally came into being.

This narrative is dedicated to Jef Frissen who was not in any way a postmodern bourgeois liberal but my father.

1 Prologue

This is a narrative about the special connections linking politics, administration and technology in what many call the 'information society'. These connections are far from straightforward as has been shown by the 'Informatisation in Public Administration' research project which began in 1986. Their significance has been a constant subject of debate among academics, in public opinion, in the world of administration and politics, and in the flourishing circuit of papers and conferences. The debate is being conducted in ambiguous terms of opportunity or threat, evolution or revolution, autonomy or control; it is still unresolved and will doubtless continue to be so for a long time.

My analysis will not put an end to this ambiguity and is indeed more likely to contribute to it. A complete understanding of the connections between politics, administration and technology is impossible because such insight would signal the end of history. This account is neither more nor less than a collection of proposals and invitations to view these connections in a particular way and to give them some form. Proposals and invitations are by their nature provisional but the reader is invited to consider the issue from the point of view of ironic postmodernism. The irony of postmodernism is that it expects that irony to apply to itself. The source of that irony is the ambiguity surrounding the subject of this narrative.

1.1 POLITICS, ADMINISTRATION, TECHNOLOGY: CONNECTIONS

Current research and debate have raised many questions about the connections between politics, administration and technology. These questions relate to issues such as the possibility of modernising public administration through technology (in particular Information and Communication Technology (ICT)); how much autonomous change is caused by ICT and the possibility of designing a strategy of controlled implementation; the impact of ICT on existing political institutions; the democratising or bureaucratising tendencies of ICT. These questions have received a wide range of answers in the literature; see Zuurmond, Huigen, Frissen, Snellen and Tops (1994) Van de Donk, Snellen and Tops (1995) and Snellen and Van de Donk (1998).

At the same time there have been many important developments within public administration and politics which are quite independent of ICT. For instance, the last few decades in the Netherlands (the main point of reference in this narrative) have seen a pattern of change, experiment and innovation which has fundamentally altered the structure and functions of the administrative system, e.g. the development of new forms of governance in response to the stubborn limits of classical intervention patterns in public administration (Bekkers, 1993). These new forms of governance are characterised by moderating ambitions for centralised steering by decentralising administrative authority, by greater interaction between governors and governed, and by the creation of formal steering procedures (Frissen, 1990). Within the structures of public administration much has been changed by the processes of privatisation, contracting out and the creation of independent agencies. The administration has thereby been differentiated. The goal is to achieve a closer correspondence between administrative organisation and differentiated, professionalised social domains.

But despite this process of bureaucratic rationalisation and differentiation, the objective of improved control over the bureaucracy and the policy domains in which it operates and intervenes has hardly, if at all, been achieved. Rather, one can see administrative structures and activities taking on the form of an archipelago, policy processes becoming increasingly circular, and relationships between administration and society developing more of a contractual than a regulated character. Self-governance is thus more often the outcome of chance than of design.

The pyramidal structure of the politico-administrative system, in which bureaucracy is the tool of a political decision-making centre, is being undermined. It is fascinating to note that this process is the result of attempts to achieve precisely the opposite; *viz.* to equip public administration to deal with social complexity more effectively with the primacy of politics as its legitimating guideline.

Technological developments demonstrate a comparable dynamic. Like bureaucracy, technology is an embodiment of functional rationality. It is still possible to sustain my earlier conclusion (Frissen, 1989), that bureaucratisation and informatisation have a *Wahlverwandschaft*. ICT is being developed and applied in order to realise the dream of control through rationalisation. Modernisation is thus still an important characteristic of our type of society.

However, recent technological developments reveal changes which, on the one hand, stem from the modernisation process but, on the other, mark a break with it. The capacity of ICT is increasing dramatically, and is turning small-scale activity and decentralisation into real technological options (cf. Negroponte, 1995). Networks are becoming increasingly important and make possible both connectivity and horizontal relationships. As a result, the mainframe is losing its importance as the dominant technological artefact and determinant of organisational logic. Virtual reality offers fantastic possibilities for the simulation and construction of realities in which the difference between representation and what is represented is blurred. The representations generated by ICT are thereby acquiring an increasingly autonomous significance.

The characteristics of ICT-developments are similar in nature to the numerous administrative changes, experiments and innovations – at least in the often unexpected outcomes to which they give rise. This similarity emerges explicitly in the connections between technology and administration.

The application of ICT in public administration is often instrumentally motivated. ICT can help to achieve the goal of modernisation. With the help of technology, services and controls can be more easily personalised. Policy ambitions are fed by the wide-scale availability of administrative information. Large-scale transaction systems coupled with intelligent systems produce profiles, discover patterns and changes, and make fine adjustments possible. The greater transparency of policy domains encourages anticipatory policies.

But at the same time ICT is also strengthening the development towards horizontalising and autonomy. The network character of social policy domains and their associated administrative configurations are supported and strengthened by electronic networks, materialising as it were a policy metaphor. However, networks have no centre, they know no hierarchy and they thereby compromise the political-administrative pyramid. Added to this, ICT lessens the relevance of 'territory' as the structuring principle of organisation: the world may become a village, but similarly, every village becomes the world. And if 'territory' loses its importance we are faced with yet another problem for the politico-administrative system which, after all, is pre-eminently based upon a notion of territory over which political power is exercised and within which social developments are legitimised ideologically by the political decision centre.

The political response to these developments seems to be chiefly one of ambivalence. The impact of ICT on politics and government is discussed largely in instrumentalist terms: it should be implemented in order to achieve economic, administrative and political objectives. It is not yet widely appreciated that these objectives are themselves changed by ICT (Ministry of Economic Affairs, 1994; Ministry of Internal Affairs, 1995). Although it may be interesting to hold a debate on the Internet about the BIOS-3 bill, which dealt with the significance of ICT for politics and public administration, one should then also address the question of whether the Internet itself does not imply another form of politics and administration.

On the other hand, administrative developments and their political significance *have* been widely discussed as, for instance in the parliamentary reflections in the light of the so-called Deetman reports. One of the central questions which they raised is whether the wide variety of forms of administrative change should be paralleled by forms of political renewal. But although the question is almost rhetorical – since a denial would suggest that political systems are ahistorical – the outcome of those reflections can not be interpreted as anything but just such a denial.

Nevertheless, two further developments do reveal some awareness of the problematic relationship between politics and administration. In the first place, there is the current advocacy of restoring the primacy of politics. It expresses a concern that the many changes, experiments and innovations have weakened political control over the bureaucracy. The political centre sees its position attacked by patterns of differentiation, horizontalisation and autonomy. The complex networks which link government and society ought to be unravelled and a vertical structure of authority should be restored.

But such political authority can only be exercised legitimately if it does not become embroiled in bureaucratic discourse and again becomes an autonomous source of meaning. This is precisely the theme of the second development: a moralistic change of direction in politics and political theory. Laying down moral values can only be authoritative if politics again dares to moralise and succeeds in formulating principles and ideals which give direction to social developments and guide the bureaucracy in its instrumental intervention patterns.

In my opinion, a political institution which finds it constantly necessary to emphasise its *raison d'être* – the existence of a public decision-making centre – is in fact in crisis. An institution's reason for existing should surely be self-evident needing only ritual reinforcement. And whether the 'recalibration' of that *raison d'être* can be found in the communitarian argument for moral politics is, quite apart from the inherent dangers of the argument itself, very doubtful.

Within the connections between politics, administration and technology, there are emerging new sets of meanings which empirically require radical change and theoretically demand a new interpretation. The empirical changes can be described as postmodernisation (Crook, Pakulski & Waters, 1992; Harvey, 1989; Poster, 1990). On the theoretical plane, postmodernism provides an interesting interpretation of the relationship between politics, administration and technology with striking consequences for the organisation and functioning of administration and politics. The disciplines of public administration and political theory will have to formulate those consequences.

1.2   POSTMODERNISATION AND POSTMODERNISM

The connections between politics, administration and technology are ambiguous. There is much talk of the advance of rationalisation and differentiation in the pursuit of better control over the politico-administrative system –

refining the bureaucratic instruments of politics – and over the social domain as a whole: perfecting the instruments of policy to achieve and stimulate desired developments.

But this modernisation strategy, as we can call these attempts at greater political control, seems to run up repeatedly against limits imposed either by social complexity or by an increasingly well-equipped bureaucracy. The strategy of modernisation thus hits the limits of modernisation itself. Some time ago (Frissen, 1989) I pointed out that the steady growth of functional rationality generates a process of differentiation whose functionality produces new sources of meaning. Patterns of meaning in the politico-administrative system move to all kinds of places within the bureaucracy and to the configurations of administrative and social actors in differing policy domains. The idea that coherent meaning is produced centrally by the political system becomes less clear-cut. The archipelago structure which emerges is not only a form of structural differentiation but also a form of cultural fragmentation.

A parallel to this within society – in the economic, social and cultural fields – is a comparable development towards differentiation in forms of organisation and behaviour. This too leads to a fragmentation of meaning because there is no longer an integrating coherence in the social processes of meaning-formation. Politics, in its ideological form, is also changing. On the one hand, we see a process of ideological generalisation: political ideologies become more abstract and uniform. On the other, it appears that politics is breaking up into all kinds of specific and often temporary patterns of ideological reconstructions similar to those observed in single-issue-movements. Ideologies do not disappear but take on a much more local significance. The crisis of the great ideologies – at least some of them – marks the failure to make of history the unfolding of a grand narrative and to design political institutions accordingly.

Finally, the belief in science as the 'legislator' (Bauman, 1992) has lost its aura of truth. Science, as an applied discipline for supporting intervention, has only been able to establish contingencies and where it has remained 'pure' it has been as a patchwork of paradigms, perspectives and metaphors. The ideal of the Enlightenment, that reason would make history a framework for liberty, emancipation and justice, has suffered a serious blow, both empirically and theoretically.

The ambiguity of modernisation is the central theme of postmodernism. Naturally this is not a unified collection of theories and opinions for that would conflict with postmodernism's self-image. It is rather a variety of perceptions and ideas where a number of core concepts of modernism are abandoned for a diagnosis of social development as non-linear, non-centralised and fragmented. The idea of steady development towards more functional differentiation on the one hand, and the realisation of Enlightenment ideals on the other, has become problematic.

Social processes are so complex that contingencies tend to dominate and any perceived patterns or connections can only be attributed to the accidental outcome of that complexity. The consequent fragmentation of meaning in turn affects the structures and processes of politics and administration since they are now no longer at the centre guiding and legitimising social developments.

Postmodernism adds a number of theoretical ideas to this diagnosis of what we might call postmodernisation. Postmodernism is anti-metaphysical: there are no grand narratives offering legitimisation, any more than there can be an *avant-garde* in the cultural domain. But if one abandons the notion of legitimising meta-narratives or *avant-garde* movements which reveal 'true' culture, what remains is just a proliferation of small, mutually related but irreducible narratives, and culture becomes a non-hierarchical collection of meanings. One is left with mass culture, an infinite reservoir of constantly changing lifestyles.

It is striking that postmodernism accords a crucial role to information and communication technology. It claims that the characteristics of the technology contribute to fragmentation, the disappearance of centres, the domination of representation and the virtualisation of reality. Though no postmodernist himself, Negroponte (1995) has shown what awaits us in this context.

Processes of postmodernisation, when we can observe them, and the theoretical position of postmodernism appear therefore to be relevant to an account of the interconnection between politics, administration and technology. I shall thus investigate the postmodern character of those connections and attempt to show that a postmodern perspective is of considerable importance for public administration and politics both in practice and as academic disciplines. The kind of perspective I have in mind is one of irony. An irony which enables one to accept that fragmentation and contingency can be meaningful.

1.3 THE NARRATIVE

This narrative is mainly about Dutch public administration. Because of the special relationship between the discipline and its subject matter, many specialists in the field have close ties with politics and administration in their own country. This stems directly from the amount of empirical research undertaken, often commissioned by the civil service itself and the years of observation required to gain an insight into the subtleties of politics and administration in a specific societal context. It is also closely related to importance of textuality and intertextuality (Van Twist, 1995) in politics and administration and thus with the importance of the mother tongue; with the desire to participate in public debates about politics and public administration. Consequently, it is difficult to speak in general terms, let alone in the dry formulae of formal language, about 'Public Administration' as an international phenomenon. I shall therefore refrain from doing so.

The special links between politics, administration and technology which I want to describe and analyse from the perspective of public administration in the Netherlands probably only exist elsewhere in Western Europe and the United States. Any relevance which my narrative may have beyond the Netherlands themselves will be subject to that proviso and all the more so because of the postmodern theoretical approach which I have adopted. In large regions of the world the relevance of these connections will not apply. A society must undergo a certain weariness and decadence – in a descriptive sense – before postmodernism can have any meaning. The rest of my narrative will reveal the importance of this *caveat*.

This narrative is the fruit of research undertaken over several years by the Tilburg-Rotterdam research project: 'Informatisation in Public Administration'. The research was conducted substantially on commission by various branches of the Civil Service. The questions formulated in the project have constantly proved to be relevant both to strategic discussions within the administration as well as to planned or spontaneous developments towards restructuring and reorientation. This study will reflect on those questions and developments in the form of proposals and invitations. The proposals are directed at the politics and administration and at political theory and public administration. The invitations aim to introduce a specific point of view: a postmodern perspective on fragmentation and interconnections in the complex reality of politics, administration and technology. Naturally, proposals can be rejected and invitations turned down. This narrative does not claim to reveal the truth. It undoubtedly contains open-ended, speculative and unsupported assertions. However, these I regard rather as figures of speech than as methodological sins.

Very briefly, this work is structured as follows: first of all, a number of recent developments within public administration are described. They can be characterised as change, experiment and innovation. The case of creating independent agencies is an example. This has brought about a change in relationships within the bureaucracy itself and in those between the bureaucracy and politics. These changes show a striking resemblance to technological developments. I describe their organisational and cultural significance and go on to point out their implications for public administration. The role of politics proves to be at issue here and I deal with various attempts to re-establish its position, – in particular, the debate on the Deetman reports and the revival of the debate on morality in politics. It will be seen that the impact of technology on the political and administrative systems has received far too little attention.

Administrative development and political debate are conducted against a background of social changes which I sketch in a variety of fragments from a perspective of modernisation and postmodernisation. The processes of modernisation and postmodernisation are developed further in an *intermezzo* where I discuss relevant insights into the phenomenon of postmodernisation and further develop the postmodern position. The work of the American philosopher, Richard Rorty, has been an important source of inspiration. After that I focus more closely on politics and public administration and attempt to show that administrative and technological developments contribute jointly to postmodernisation. This leads to a questioning of the role of politics. Its significance for politics and administration in both practice and theory forms the central theme of the final chapters of the narrative. There I consider the theory and practice of postmodern public administration and politics. The acknowledgement of fragmentation and contingency is central. The necessity for public irony is the inspirational motif.

2 Public Administration

2.1 CHANGE, EXPERIMENT, INNOVATION: AN IMPRESSION

Many narratives are told in public administration. Narratives which give expression to interests, insights and passion. Narratives which arise from a political position or reveal a bureaucratic perspective. Narratives which plead for normative readjustment or technological improvement. One narrator speaks of reassessing the welfare state (Godfroij & Nelissen, 1993: 16); another of the popularity of 'bureaucrat-bashing' (Ringeling, 1993: 6); yet another talks of strategic re-orientation (Naschold, 1993: 17–19). It is clear that, as so often in the past, public administration is again under discussion and yet again the debate is about the legitimacy of state intervention. (See also: Bekke & Kuypers, 1990; Kooiman, 1993.)

We can, however, set a different tone for the narrative. It does not necessarily have to suggest an atmosphere of crisis and decline. It is equally possible to strike a positive, generous, even enthusiastic note when talking about developments in public administration over the past decades. And then one is struck in particular by change, experiment and innovation. A few examples will illustrate this. (See Frissen, 1990: 19–20.)

Steering at a Distance

In educational policy the idea has gained ground that professional bodies should be allowed to manage themselves. Less intensive and more global regulation should make this possible and ensure that institutions are accorded more room for autonomous policy-making. However, for the various subdivisions within educational management, a collaborative model is stipulated. In higher education, this means a model of 'planning through dialogue', whereby educational establishments and government attempt to reach consensus (Snellen, 1987).

Steering by Incentive

In many areas of state-governance, incentives are used as an alternative to rules and regulations to stimulate desirable developments or performance. There is a free choice whether or not to bring about the ends desired by the government. Consent leads to financial reward, but refusal does not face sanctions.

Governance Based on Output and Performance Indicators

In various domains, the government is reluctant to be over-prescriptive about the design and production-style of social institutions that provide services. Governance and, where appropriate, financing, focus on the level and quality of output, which is tied to specific criteria. The manner in which the output is achieved is the responsibility of the institutions.

Network Governance

In the field of urban development, (local) government operates increasingly through networks of different actors involved in areas of government concern. In cases of large-scale development projects requiring considerable financial investment, the phenomenon of 'public–private partnership' is on the increase. In this framework, the government enters into a contractual relationship (sometimes, though not always, as a private legal person) with private companies or other organisations. Here one can no longer speak of a principal-agent relationship but rather an equal, collaborative association in which the content of the project itself forms part of the collaboration (see Tops, 1994; Depla & Monasch, 1994).

Independent Agencies

Tasks carried out by government organisations can also be carried out 'at a distance'. The most extreme form of this is privatisation, where a government task is hived off and taken over by a private actor. Less far-reaching forms do not involve handing government tasks over to the private sector, but to more or less independent public agencies or quangos. We shall explore this more fully later (§2.5 and §2.6).

Deregulation

New legislation and revisions of existing legislation seek to reduce and simplify rules and regulations. An advisory review committee for legislative projects makes recommendations intended to improve the quality of legislation in terms of the relationships between government and society.

Decentralisation

For some time now, decentralisation has been widely propagated as a means of reducing the burden of state governance and bringing policy-making processes closer to the citizen. To achieve this, tasks are transferred to the provinces and districts, while central government confines itself to a supervisory role – often in order to guarantee some degree of uniformity in policy-implementation.

Reorganisation

The structure of public administration is itself undergoing change, on the one hand through many forms of functional regionalisation, like policing (see: Gooren & De Zwaan, 1993) and on the other through modifications of domestic administration, as in the creation of regions out of the larger conurbations. It is striking that reorganisations occur only after decades of fruitless debate and bookshelves full of neglected proposals.

Business Methods

Within government organisations a host of initiatives are undertaken to increase functional efficiency. Self-management and contract management are the most obvious examples. Some Dutch cities have even made a feature of this in their publicity, believing that it will give them the edge in competition with international rivals. (See also: Tops, 1994: 157 ff.)

2.2 THE NEW GOVERNANCE: CHARACTERISTICS

Change, experiment, innovation. A narrative about public administration can with some justification be set in such terms. In his study of future trends, In 't Veld even marks out 1993 as a turning-point in the debate on the organisation of the civil service (In 't Veld, 1993a: 14). The Netherlands is by no means unique in this respect. Naschold (1993) describes comparable tendencies in various OECD countries.

More generally, change, experiment and innovation can be seen as indicators of attempts to institutionalise new forms of governance by and within public administration. These new forms of governance have emerged during recent decades in all areas of state intervention as an alternative to classical forms of steering, but which nevertheless build on specific traditions in Dutch public administration and their associated political ideologies. In the Anglo-Saxon world, a term often used for this is 'New Public Management'. In spite of all their variety, these new forms of governance share certain common characteristics (Frissen, 1990: 20-21).

The Extent of Governance

Governance is becoming more global and steering, as such, less intensive. This is a negative demarcation: government intervention is being reduced in favour of processes of social self-regulation.

The Object of Governance

The number of application points for direct steering is diminishing. In theoretical terminology, state governance or steering, depending on the specific policy domain, is either being limited to what Snellen (1987) calls 'governance by key parameters', (budgeting on the input side, incentives on the output side), or it is taking on the form of meta-steering. In the latter case, it focuses on the decision-making arrangements in a particular policy sector, and no longer on the actual content of the decision-making. Government thus confines itself to the structures and procedures of decision-making and is, in a sense, indifferent to their outcome (Snellen, 1987:21–25).

The Level of Governance

Steering responsibilities are being differentially situated. The one-sided emphasis on the macro-level is disappearing in favour of a re-evaluation of governance at the intermediate and micro-levels of society. This involves social organisations and interest groups as much as provincial and local authorities.

Participation in Governance

The empirical observation that social steering-processes occur in networks of organisations and individuals becomes the starting point for designing decision-making arrangements. This multi-actor perspective means that communal responsibility for steering is incorporated into decision-making arrangements.

Perspective on Governance

Instead of the traditional centralised model of control, based on a rational organisation which plans and manages from a single point, we have a model which treats differentiation, variety and multiformity as basic principles rather than obstacles to be overcome. A top-down approach is replaced by a bottom-up approach and control is decentralised in networks of organisations and actors. Central responsibilities are subsequently defined as derivatives of this.

2.3 THE ANTIPODE: CLASSICAL GOVERNANCE

The antipode of these new forms of governance is so-called 'classical governance'. This too, of course, is a generalised term for an ideal type, and refers to the forms of governance associated with the heyday of the welfare state and a belief in social engineering. That belief also provided the politico-ideological legitimisation of classical governance.

Classical governance, in theory and practice, often takes the form of planning, a term which appears to have disappeared entirely from contemporary discourse, and, according to Bekkers (1993), is a voluntaristic and constructivist activity: 'Planning as an instrument of governance is viewed as a perspective on, and a programme of, long-term integrated social renewal' (Bekkers, 1993: 33).

Following Bekkers (1993:31–58) we can distinguish the following characteristics of classical governance:

Structure

Public administration is highly mechanistic: it functions like a machine, 'the machinery of government', and any problems in steering it can be cured through fine-tuning and adjustment. Moreover, steering the machine is a strongly top-down activity within a hierarchical pyramid with 'a central and sovereign centre of governance' (Bekkers, 1993: 39). The functioning of public administration is given shape by programmes, which are constructed like a knowledge-tree – a cognitive structure of hierarchically arranged goals and means. According to Bekkers, classical governance is based on '... the postulate that an organisation can be effectively steered if the structure of the steering organisation is set up like a rationally designed machine with clear lines of authority and where tasks are built on each other logically and as an extension of each other' (Bekkers, 1993: 40). (See also Frissen, 1992a: 48–51.)

Application Points and Context

The object of governance is fully known and, at the same time, loyal to its steering body. It is this which makes it possible to plan and manage social processes. Governance is exercised from a single archimedian point situated outside and above society. In technical terminology, there is a steering organ which steers the steered system by means of steering signals. This dichotomy also explains the steering centralism of classical governance (see Den Hoed, Salet & Van der Sluijs, 1983: 43.). Every aspect of the internal and external workings of the steered system are, in principle, application points. And the steered system can constitute the whole of society. Thus, in classical governance, context and application point are one.

Knowledge and Information

Rationalisation is the most important goal of classical governance. 'To know more is to steer better', as the adage has it. Knowledge-acquisition and information-gathering are organised as systematically as possible. The knowledge is often formulated as policy theory: a systematic whole of causal, final and normative relations within the steering domain. Many techniques and methods for improving rational governance have been and are still being developed.

Steering Modalities and Steering Instruments

In classical governance the 'command and control' approach predominates. Control is an important ambition, and planning is one of the crucial steering instruments of classical governance. The megalomaniac PPBS is a well-known example. Classical governance is also strongly rule-oriented. Through direct regulation, attempts are made to change society: modification through influencing behaviour. An instrumentalist perspective on legislation and regulation is predominant.

It is in opposition to the classical governance of the welfare state – the interventionist state – that the new forms of governance offer an alternative. On the one hand, classical governance has limitations and suffers from considerable inefficiency and ineffectiveness; on the other hand, objections of principle have been brought against it. There is, for instance, the neo-conservative criticism of the welfare state to which Reagan and Thatcher gave concrete expression in their administrative reforms. But there has also been a relative decline in classical governance which is reflected in new forms of steering in a variety of administrative domains.

2.4 THE TRADITION: SOVEREIGNTY, SUBSIDIARITY, CORPORATISM

In a number of respects, the new forms of steering are not at all new. They build on particular traditions in Dutch public administration and their associated political ideologies. In the first place, Dutch public administration is characterised by a history of multiformity and coalition-formation (see Van Holthoorn, 1988). Its multiformity is rooted in the philosophical and religious differentiation of the Dutch population, and reflected in its institutional 'pillars' and multi-party political system. Because of this multiformity, there has always been a need to form political and social coalitions and this in itself has acted as a brake on any excessive centralisation by the administration. Since no single interest group can obtain a majority, even politicians have to acknowledge the need for incrementalism as the basis for governance. At the same time, participation in steering processes by different and differing social actors then becomes an intelligent model for both incrementalism and multiformity. This combination of multiformity and coalition-formation is strikingly described by Lijphart (1968).

Organisationally, we can also recognise these aspects of multiformity and coalition-formation in such phenomena as functional decentralisation, deconcentration, independent administrative organs and agencies. In the socio-economic arena the Netherlands has a strong tradition of corporatism and neo-corporatism, recognisable in the tripartite model adopted for the privatisation of the Department of Employment. (See Frissen, Albers, Bekkers, Huigen, Schmitt, Thaens & de Zwaan, 1992: 11.) The participation of citizens and interest groups, partnerships in the steering of social processes, networks and principles of self-organisation and self-regulation are all modern outcomes of a long tradition of (neo)-corporatist administrative arrangements. (See also: Verhallen, Fernhout & Visser, 1980.)

Both neo-corporatism and the involvement of social organisations (the 'midfield') in forming and implementing government policy are traditional predecessors of the new forms of governance discussed earlier. The ideological inspiration for this tradition – the Christian-Democratic idea of subsidiarity and 'the sovereignty of the social unit' – turns out to be surprisingly topical, as Hirsch Ballin (1988) rightly argues.

Catholic thinking on the state and society was originally corporatist in character and as such conflicted with the protestant ideology of 'sovereignty of the social unit' or 'unit sovereignty' which postulates a fundamental independence and impenetrability of discrete social spheres (Hirsch Ballin, 1988: 120–121). Corporatism, in contrast, assumes 'a political competence which encompasses the whole of secular social existence ... a concept which is also reflected in the description of the state as *societas completa*' (Hirsch Ballin, 1988: 122).

A close affinity exists, however, between the concept of 'unit sovereignty' and the principle of subsidiarity. The relationship between the two is the idea of dispersed responsibilities and competencies within society. Unit sovereignty emphasises the autonomy of differentiated spheres of life; subsidiarity emphasises the complementary nature of political, public and private competencies. Autonomy indicates boundaries; complementarity implies an obligation to intervene 'in the material and non-material needs of society' (Hirsch Ballin, 1988: 125).

Much of the current content of new governance which gives expression both to the autonomy of social domains and to the need for participation, communication and partnerships, is thus recognisable in the ideological tradition. Perhaps it is this combination of continuity and discontinuity, of tradition and renewal, which explains why new governance has attracted such wide interest within Dutch public administration.

2.5 A CASE: INDEPENDENT AGENCIES

Independent agencies are a striking illustration of the combination of tradition and renewal. The independent agency is a new form of governance and fits into the process of change, experiment and innovation. In that sense, it can be seen as one of the attempts to modernise public administration. At the same time, it revives the tradition of subsidiarity, unit sovereignty and corporatism, because while attempting to autonomise, it also lays down new, often horizontal, connections. The independent agency is not new: in the Netherlands we are familiar with hundreds of independent administrative organs. But what is new is the revival of interest and its centrality in the current political and public debate about the organisation of the civil service (See also: Frissen *et al.*, 1992a; Frissen, Hirsch Ballin, Hoekstra, Visser, Kuiper & De Boer, 1993; Kickert, Mol & Sorber, 1992; De Vries & Korsten, 1992.) An independent agency is an organisation that implements public tasks. The organisation can be public as well as private. As an agency it is part of the public sector and its formal position has a constitutional basis. It differs from ordinary governmental organisations, because ministerial responsibility is limited to those competencies laid down in the legislation which is always needed to create an independent agency.

Forms

Independent government organisations take a wide variety of forms. The following may be distinguished (see also: Frissen *et al.*, 1993: 17–18):

a. Internal independence by differentiation and control. This is generally a matter of self-administration and may be combined with contract management. The self-managing unit has autonomous powers in respect of human and material resources. It is answerable for the achievement of its terms of reference: its output. Task definition and budget are regularly defined in a management contract between departmental heads and the self-managing unit.
b. Internal independence through decentralisation or agencies. Agencies combine self-management and decentralisation. Parts of the organisation are administratively, and often physically or geographically, self-sufficient and are given self-managing autonomy. This autonomy relates to resources (including conditions of employment), enhanced end-of-year flexibility in investment projects, control of reserves and capital.
c. External independence through functional decentralisation. Functional decentralisation has a long tradition in Dutch public administration, and functionally independent administrative organs are particularly relevant to the wider issue of independent agencies. Boxum, De Ridder and Scheltema (1989) distinguish the following types:

- the specialist-type, concerned with tasks requiring specific expertise.
- the independent authority-type, whereby far-reaching programmable decisions are taken independently of the government but under political influence.
- the participation-type which carries out complex and inclusive tasks requiring systematic consideration, and involving both experts and interested parties.

d. Independence through privatisation. Privatisation involves transferring responsibility for ensuring that facilities exist or are available, or for their production and distribution to the private sector. Both termination of policy and subcontracting fall under privatisation.

Motives

There are various motives for granting independence to government organisations. They can be summarised under two heads (see also Frissen *et al.*, 1993: 18–19).

Politico-ideological motives

- Independence results from a discussion about core tasks. The government should limit itself to these core tasks. Remaining tasks can then be carried out by independent organisations or handed over to the private sector.
- Independence contributes to the strengthening of political primacy. Through independence the process of political decision-making is relieved of concern for detail and can focus on the broader issues. In this way, political primacy can be strengthened if not entirely regained.
- Independence, especially in the form of functional decentralisation, strengthens the democratic content of public administration, argues Scheltema (1977). Through independence, the structure of public administration becomes more transparent, which facilitates control and enhances openness. Citizen participation can be increased and the distribution of power can be improved.

Instrumental-professional motives

- Independence leads to a greater awareness of costs and improved task-orientation. Independence increases flexibility and enhances job-satisfaction while also being conducive to 'payment by results'.
- Independence for parts of the policy-making cycle makes more specific management of discrete phases possible. This becomes

necessary because of the unique characteristics of policy-formulation, policy-decisions and policy-implementation.
- Independence makes it easier to take into account considerations of scale, bringing in specific expertise and the participation of interested parties (Boxum *et al.*, 1989).
- The decentralised autonomy achieved through independence is an adequate response to complexity, professionalism and turbulent policy domains. The constraints on central steering can thereby be lessened.

Types

In a study of independent agencies undertaken by our research group, two ideal-typical configurations were formulated on the basis of theoretical observation and the empirical analysis of sixteen cases. The characteristics of the policy-making process in the domain where the independent agency was created, turned out to be definitive. I shall discuss both types (see also Frissen *et al.*, 1993: 21–23).

Policy-making as sequence
In the first type, the policy-making process can be conceived of as sequential. The separate phases of the process – preparation, development, determination, implementation, evaluation, feedback – are conceptually and empirically clearly distinguishable and linked sequentially. The design of the process conforms entirely to the classical conception of the policy-cycle. In this case policy can be presented within a cognitive structure of ends and means whereby the means are hierarchically subordinated to the goals.

Such a clear demarcation of phases is only possible if policy-making and the policy domain satisfy a number of conditions:

- The policy objectives are defined by a central actor
- The policy is relatively complete: ends and means are established and uncontroversial.
- The conditions under which the policy is to be implemented are highly stable.
- There is a clear framework of ends and means.

If the policy-making process has these characteristics, it is possible to give independence to the parts of the organisation that deal with specific phases of the process. As a rule, these have to do with operational implementation, although supervisory or evaluative functions may also be considered.

From the point of view of the organisational structure, this form of independence is often described as functional, whereby hiving off specific elements leads to benefits of scale and standardisation.

Steering the independent unit occurs mainly at the input and output stages. Information systems can usually supply quantified information about the input and output parameters. Typical examples are: National Land Survey, Government Computer Centre, Department of Transport, Student Finance, Inland Revenue.

Policy as interdependency
The second type is in every respect the converse of the sequential type. The separate phases of the policy process are neither conceptually nor practically distinguishable, and there is a high degree of mutual interdependence. They are not activated successively and often take place simultaneously. The process has a network character; it is an arena of actors with frequently competing interests. This form of policy-making can not readily be conceived of as a cognitive structure of goals hierarchically connected to means.

The policy process for a specific policy-domain has strong interdependencies when:

- the policy goals are set by differing and often conflicting actors;
- the policy is incomplete and often controversial in respect of its aims and the means to be employed;
- the environment of the policy domain is dynamic and lacks stable conditions for policy implementation;
- there is no clear framework of ends and means.

When the policy process has these characteristics, it remains possible to make parts of the organisation independent. Indeed, perhaps there is an even greater necessity to do so. But the independence then affects the whole policy cycle: wider responsibility for all the policy-phases is now decentralised. In addition, we sometimes see in practice that under these conditions even 'sensitive' phases such as allocation or evaluation and supervision are also made independent.

The organisational structure of this type of independent unit is usually market-oriented because whole areas or sectors are being served.

The steering of independent units occurs at the points of interdependence with their environment. Modalities of steering are, then, the structuring and procedurisation of decision-making arrangements, or the furthering of dialogue with the community (Snellen, 1987). The provision of information is most intense between policy-phases, and between organisations and environment. The information is usually highly qualitative and often disputed.

Examples are: Central and Regional Employment Offices, the Central Office for Medical Provision and the Commission for the Media.

2.6 CASE 2

In the current debate about the creation of independent agencies, two sets of opinions are aired most frequently. Firstly, independence is seen as a logical outcome of the operation to limit the activities of central government to core tasks. Secondly, one continually hears the view that independence should be given to those parts of the organisation most directly involved in policy implementation. One could take issue with either view. Furthermore, reservations might be expressed about the consequences of independent agencies for political primacy as well as about the popular concept of core departments.

Core Tasks

The first point about the debate over core tasks is that it is extremely difficult to distinguish between what is core and what is subsidiary. It seems that core tasks can not be objectively defined but depend largely on ideological preference. Furthermore, government tasks are closely interwoven through the links between legislation, administrative practice, systems and operating procedures (see Hirsch Ballin, 1991). Secondly, in practice, core tasks themselves are considered as candidates for independence, both in already existing forms as well as future ones under discussion. Examples are the executive organisations in the field of social security and the proposed privatisation of prisons. In all these cases, tasks are involved which many consider to be crucial to government functions and are sometimes integral to the classical functions of government. From the foregoing it can be seen that neither conceptually nor practically can the distinction between core and subsidiary tasks offer a basis for decisions on independence.

Policy and Implementation

The second point relates to the distinction drawn between policy development and definition on the one hand, and policy implementation on the other. There are both empirical and theoretical objections to this distinction. In actual practice, tasks other than policy implementation have been or are being made independent. The Offices for Data Registration and Insurance are responsible for monitoring or evaluation; allocatory functions, too, are performed by independent organisations, like the Commission for the Media and the Central Office for Health Care Tariffs. Finally, in practice, the entire policy-making cycle (from policy development to implementation) can be made independent as in the case of the Central Board for Employment.

Theoretical objections to separating policy development from policy implementation have emerged from a range of studies of the subject. They show that implementation is not merely a mechanical process generated by earlier phases of development and definition but that development and redefinition also take place during implementation.

This means that in many policy domains, the organisational separation of development from implementation is unrealistic.

The Primacy of Politics

It is often argued, especially in light of the discussion about core tasks, that independence contributes to a strengthening of political primacy. Apolitical administrative tasks are separated off so that politics can concern itself primarily with the main lines of policy. This argument is only partially valid. Because independence changes the hierarchical relations between the senior political officials of an organisation and its independent parts, political primacy is also affected. Constitutionally it may simply be reflected in a change of ministerial responsibilities so that in formal terms only external independence appears to be significant for political primacy. In actual fact, one can speak of three possible changes in the primacy of politics (see also Frissen *et al.*,1992a: 74-75).

Curtailing political primacy
In the first place, creating independent agencies curtails political primacy because certain phases of the policy-making process – usually implementation, though sometimes supervision or quality control – are no longer subject to direct political governance or control, and obtain a significant degree of professional autonomy. Secondly, political primacy is curtailed because independent agencies – as a rule, the interdependency-type – involve non-political actors in decision-making. In the third place, political primacy may be curtailed as a result of deliberate choice. New governance – of which independent agencies are an example – frequently express a decision to make government step back.

Enhancing political primacy
Independent agencies can lead to more effective political primacy if they allow political decision-making to concentrate on basic principles, on the formulation of strategy. It is freed from the need to intervene in and control the detail of administration.

Reorientation of political primacy
Independent agencies leads to a re-orientation of political primacy to the extent that politics becomes a question of meta-steering, laying down standards and safeguarding decentralised administrative practices. By decentralising administration and creating networks of different, including non-political, actors, the actual content of the objectives laid down by the central political actors (government, parliament) becomes less important.

Other tasks, however, like meta-governance, standards and safeguards become more important. We shall return to this later.

Core Departments

In the current debate over administrative independence, the reasoning often goes as follows: central government must be quantitatively reduced in size and qualitatively confined to core tasks; to this end, the implementation of policy can best be entrusted to independent administrative organisations which maintain a more or less loose connection with the political–bureaucratic leadership; departments then become core departments with mainly administrative and strategic responsibilities. The Permanent Secretaries' report on the organisation and procedures of the civil service and the report of the Wiegel Committee (Commissie-Wiegel, 1993) both give credence to this line of reasoning. The Permanent Secretaries' report, in particular, appears to favour clarity and unambiguity, and a desire to disentangle, reduce complexity and separate out functions wherever possible.

> The trend [towards results, effectiveness and efficiency] implies that there is clarity about the goals of government action, about the effectiveness of government action and about the accountability for results of the government organisations involved. It compels a sharper separation of responsibilities between the public and private sector and a more careful assessment *ex ante* of the costs of decision-making and implementation. It requires a breaking down of government action and selective restrictions in the development of policy. (Ministerie van Binnenlandse Zaken, 1993: 4-9)

This echoes a preference for purification rituals (Van Gunsteren, 1984) and dichotomous thinking. The Wiegel Committee was more clearly inspired by a recognition of reflexivity as a fundamental distinguishing feature of social systems. It is reflected in their wider interest in the horizontal and vertical links between core departments, administrative organisations and the social environment.

Nevertheless, some important problems remain :

- The separation of policy-making from implementation is perhaps organisationally possible , but conceptually only under clearly stipulated conditions (stability, consensus, a complete policy).
- The creation of core departments confirms the illusion that centralised strategy-formulation in complex and professionalised environments is possible.
- Core departments will rapidly begin to formulate pretty blueprints for the future, or present trivialities in the guise of policy.

- Independent executive organisations responsible for policy implementation will claim an authentic position in the policy-making process and if this is not obtained, pursue autonomous policies.

These observations arise from this particular case and form an introduction to the following paragraphs which sketch in more general terms the significance of change, experiment and innovation in public administration.

2.7　AMENDMENTS TO BUREAUCRACY

Many forms of new governance, much change, experiment and innovation can be regarded as amendments to bureaucracy. They aim to formulate and generate solutions for problems in bureaucratic functions. In this context, bureaucracy is conceived of as an organisational pattern which is hierarchical in nature, which functions on the basis of specialisation, standardisation and formalisation, and 'loyally' – i.e. *sine ira et studio* – implements political preferences. Bureaucracy is an important part of the modernisation process in our type of society. It embodies advancing functional rationality and places it at the service of political and legal domination (Weber, 1985 (1922); Zijderveld, 1983; Frissen, 1989). In the words of Weber:

> Experience tends universally to show that the purely bureaucratic type of administrative organisation - that is, the monocratic variety of bureaucracy - is, from a purely technical point of view, capable of attaining the highest degree of efficiency and is in this sense formally the most rational known means of exercising authority over human beings. It is superior to any other form in precision, in stability, in the stringency of its discipline, and in its reliability. It thus makes possible a particularly high degree of calculability of results for the heads of the organisation and for those acting in relation to it. It is finally superior both in intensive efficiency and in the scope of its operations, and is formally capable of application to all kinds of administrative tasks. (Weber M., 1968: vol. I, p. 223)

This characterisation of bureaucracy is, of course, an ideal stereotype: in actual reality there are variations, particularities and nuances. Nevertheless, many of the ideal-typical characteristics of the sociological concept of bureaucracy have always been empirically realistic for government bureaucracy. Hierarchy and loyalty, specialisation and centralisation, standardisation and formalisation are all still important features of many government bureaucracies. However, there exist a number of problems, even pathologies connected with these features, and many experiments and innovations are directed precisely to their cure: removing dysfunction and refining and perfecting the bureaucracy. I shall mention a number of them.

Reorganisation

Over the past decades there have been innumerable attempts to tinker with government organisation. I shall confine myself to the level of the State Civil Service but just as many examples can also be observed at provincial and especially district level.

A background survey conducted for the Wiegel committee and published in its report (Ministerie van Binnenlandse Zaken, 1993: 72-113) provides a useful summary of the attempts made to adapt central bureaucracy to changing circumstances and to modernise it. The study includes a helpful table, which is reproduced below.

These large-scale reorganisations are naturally supplemented, expanded and alternated with the minor reorganisations which continually take place in the administration.

Next to them, there are the major reorganisations of large sections of the administration such as the police, the judicial system, employment provision, various departments etc.

Reorganisation can be seen as an attempt to repair the machinery, modify existing structures and functions, improve coherence and flexibility. Reorganisations are an important way of giving meaning through their rhetorical qualities: they appeal to the techno-scientific rationality of planned change (March & Olsen, 1983: 282–284). In that sense, they satisfy the need for intervention to cure dysfunction. As the table of reorganisations shows, the aim is always improvement and change at the same time. On the one hand, what exists must be optimised; on the other, its functionality must change. Less overload and improved tuning by redistributing tasks; less compartmentalisation and more integration through a different distribution of policy domains; reducing conflict and improving flexibility by changing the system of conflict regulation; reducing inefficiency through independent units. All are amendments to the narrative of bureaucracy.

Modifying the Toolbox

The functioning of bureaucracy is also modified through changes in the tools which accompany new forms of policy steering. De Bruijn and Ten Heuvelhof believe that much of the renewed interest in steering instruments is related to growing concern about the steering capacity of government (De Bruijn & Ten Heuvelhof, 1991: 3). They describe the new instruments as 'second-generation', one of whose characteristics is the way they discount any limitations or barriers to steering, both conceptually and in practice. In other words, the barriers are recognised and acknowledged and then integrated as conditions into the instruments of governance. These barriers are seen as 'social multiformity or differentiation, the closed and autonomous

Table 2.1: Reorganisation of the Civil Service

Period	1972–77	1979–1981	1982–86	1987–1991
Most important source	Van Veen Commission, the WRR, the Mitaco	Vonhoff Commission (CHR)	Head of the Civil Service	'rethink' workgroups
Object of research	interdepartmental task distribution and co-ordination	main structure and functioning of the Civil Service	networks in and around the public administration	organisation and management of government
Under investigation were	actual complaints and bottlenecks	factors contributing to government controls more appropriate to the steering needs of society	(faulty) links between (government) actors and the rules governing these links	possibility of more businesslike management by government
Chosen approach	practical approach	systems approach	network approach	businesslike approach
Concept of government	admin. apparatus with vertically organised departments and horizontal policy-forming organisation	a composite whole, an open system interacting with its environment	break up into different inter-organisational networks in which all actors negotiate on several fronts at once	organisation which produces goods and services for which it must sacrifice scarce resources
Problem analysis	departmental divisions and interdepartmental co-ordination unsatisfactory and an overloaded top.	the powerful world of the sectors dominates the weak world of integration: the result is compartmentalisation	the rules of the game are unclear so that office politics can lead to fatal conflicts and rivalries	insufficient room for powers and responsibilities at the local level
Chief objective	improve harmonisation	increase integration	increase flexibility	increase efficiency
Solution type	co-ordination model	main policy areas	conflict adjustment	independence
Implementation strategy	substantiated model proposal	set up a steering group and project organisation	incremental change	work with pilot schemes
Reason for report	scientific research	political pressure from above	decentralised support	tests of practical value

Source: Netherlands Ministry of Home Affairs, 1993: 104

nature of the actors to be steered, and the interdependencies between actors, both public and private.' (De Bruijn & Ten Heuvelhof, 1991: 13). They list the following instruments:

Multilateral instruments
These are forms of agreement between different parties. They can relate to concrete and discrete matters or to meta-agreements and agreements about institutional formation (De Bruijn & Ten Heuvelhof, 1991: 81ff.).

Person-orientated instruments
These include the steering agent and the use of personal networks. Here the loyalty of the steering agent, the support provided, and the regulation of responsibilities are of importance (De Bruijn & Ten Heuvelhof, 1991: 97ff.).

Incentives
Desirable actions are rewarded, undesirable actions are penalised. The incentives and disincentives are frequently financial (De Bruijn & Ten Heuvelhof, 1991: 113ff.).

Indicators
The output or input of the organisation to be steered are tied to specified indicators. Indicators are generally quantified (De Bruijn & Ten Heuvelhof, 1991: 129ff.).

Communicative instruments
These include information and propaganda as well as various forms of interactive information-exchange (De Bruijn & Ten Heuvelhof, 1991: 141ff.).

The modified steering toolbox - also an example of change, experiment and innovation - is equally an amendment of bureaucracy. The classical bureaucratic activities of steering and control are not relinquished. Rather, they are made more realistic, more attuned to the stubborn reality of limits and barriers. At the same time, the process of 'perfecting' bureaucracy continues by making steering and control more sophisticated. And yet there is also change, the core of which lies particularly in the fundamental recognition of limits and barriers to governmental steering.

Business Methods

Apart from changes in structure and the available instruments, there is also a change in discourse. During past decades, a business culture has been introduced into public administration. Where previously a dichotomy had always been assumed to exist between market and hierarchy, between business and government, between private and public, the recent past has seen the adoption of business practices as the norm for government and governmental action. In applying that norm we come across a range of new principles: from the rhetoric of the superiority of the market to market-oriented rationalisation.

Demands that many government activities should be shed or merely abandoned to the market have not met with much political support in the Netherlands. More popular has been an approach which expects government to be more 'business-like'.

The business-like approach is to be seen in:

- the use of market-oriented tools, both in the form of incentives and the introduction of the profit motive
- the development of an 'entrepreneurial attitude' in policy formation and implementation, through, for instance, an orientation towards 'partnerships', 'stakeholders' and agencies.
- the stimulation of a client-oriented culture in those parts of the bureaucracy responsible for public services.
- the encouragement of keener financial and economic awareness at all levels of the government apparatus.

This reflects a change of ideology and idiom, the real effects of which can be observed all over the world irrespective of the political orientation of governments concerned (Crook *et al.*, 1992: 99–101). At the same time, the change has not affected the foundations of bureaucracy.

Modernising Bureaucracy: Rationalisation

The changes described above – reorganisation, modified toolbox, and business methods – are evidence of amendments to bureaucracy. They do not fundamentally change the pattern; rather, they adapt or refine it. It is, in other words, a modernisation or further rationalisation of bureaucracy. The most important goals are increased effectiveness and efficiency; the most important results are streamlining and refinement. Undeniably, financial problems have also played an important part in this. Pressure on government expenditure has necessitated reorganisation and streamlining and has resulted in more intensive attention for financial and economic management at both the macro- and the micro-level.

Next to this, there has been a more business-like approach to the government's service role. The paternalistic attitude of the welfare state has been criticised. Rationalising in the sense of demystifying the quasi-religious aspects of the 'caring state' has led to a reorientation of state tasks and of the relationship between state and citizen.

Finally, there is talk of further rationalisation in the structure and functions of public administration. Reorganising and adapting steering tools should contribute to more effective government action. The governmental apparatus needs to be redesigned to fit in with developments and parties in policy domains. The steering tools must complement this by being adapted to respond more flexibly to conditions of increasing complexity. To use Tops' phrase (Tops *et al.,* 1994), there is a strategy of modernisation aimed at increasing decisiveness and vigour. In sociological terms, it is 'classical' modernisation directed at the further rationalisation of public administration.

And yet there is more to it than that.

2.8 DIFFERENTIATING THE BUREAUCRACY

Change, experiment and innovation in public administration are not only amendments to the bureaucracy. They can also be seen as a steadily increasing differentiation of bureaucratic organisation patterns. Differentiation must here be seen as refinement and particularising in the sense of greater variety and multiformity in order to operate with greater flexibility. One can observe three developments: localisation, functionalisation and autonomisation.

Localisation

Within public administration and in a variety of areas of policy-making, localisation is an important development. (See Frissen, 1991: 20) It is a matter of strengthening local elements within an organisational context or strengthening local orientation within a policy domain or policy cycle. Decision-making powers are linked directly to the parts closest to the citizen or which can most rapidly bring policy development in line with social developments and problems. Decentralisation is desirable in such situations in order to bring about greater flexibility. (See too Godfroij & Nelissen, 1993: 509)

A similar tendency can be observed in the business world where local components (subsidiaries, business units, profit centres) are given greater decision-making powers so that they can respond more flexibly to changes in the market. (See Clegg, 1990: 176ff; Harvey, 1989: 141ff.; Peters 1992: 129ff.) Similarly we can see a strengthening of local authorities in public administration. In particular, local authorities are taking over more tasks in such policy domains as social renewal, welfare, education and the urban environment. Underlying this is the conviction that local authorities are 'closer to the citizens', and can thus contribute to more responsive government.

Internationally, the tendency to localisation is reflected in the increasing importance of regions. Regionalisation leads to a relative decline at the national level. Again, the reasoning is that the region is a more appropriate level because it is closer to the relevant policy problems and the actors involved.

Functionalisation

Another tendency which can be observed in a number of policy domains is the strengthening of the functional dimension in policy making. Functionalisation has long been a feature of Dutch public administration (see §2.4) in its habit of giving administrative authority to involved or interested parties and experts. Functional administration, often in decentralised form, is an alternative to territorial administration. Its authority relates to a specific policy domain, not a geographical area. Administrative powers are often exercised by autonomous administrative bodies which carry out their statutory duties independently of ministerial supervision. In many cases, policy responsibilities are delegated to the social 'mid-field'. Recently, functionalisation has been combined with regionalisation in the creation of functional regions. For specific tasks – police, the fire service, transport, employment etc. – regions are created and managed by involved and interested parties and/or experts.

At the same time, functionalisation can be seen as a form of administrative independence. This tendency has intensified in recent years leading to criticisms that the Netherlands is becoming an administrative patchwork. The counter-movement – certainly in respect of functional regionalisation – is the creation of a geographical administrative layer at the regional level, as in the case of the larger conurbations.

Autonomisation

Partly connected to the developments discussed above is a trend towards autonomising organisational sub-divisions and social domains. On the one hand, this occurs through converting parts of the governmental organisation into independent agencies (see §2.5 & §2.6); on the other hand, one can speak of autonomisation if more room is left within policy domains for self-steering and self-regulation. (See Eijlander, Gilhuis & Peters, 1993.)

In the creation of independent agencies, a functional administrative level is generally created which functions more autonomously in respect of the departmental (or district or provincial) organisation. Autonomy provides advantages of greater flexibility. It also reduces political responsibility which frees the body concerned from the strait-jacket of central steering and control.

Self-steering and self-regulation increase the autonomy of a specific social domain. That means less central regulation, but more particularly a reorientation of regulation and a shift of regulatory responsibility to other social contexts. At the central level there is global regulation and normative frame-setting whereas locally the norms are substantively specified. Autonomisation is thus a tendency which, as with independent agencies, self-regulation and self-steering, leads to further differentiation within the public administration and more generally in the complex regulation and governance of society. (See also Kickert, 1991:18–19.)

Modernising the Bureaucracy: Differentiation

Localising, functionalising and autonomising are forms of change, experiment and innovation which are linked in the acknowledgement of social multiformity and variety. That acknowledgement means that variety and multiformity are recognised in the first place as relevant social characteristics and secondly are taken as a basic principle for administrative action and are not perceived as problems to be overcome. Steering units and the steering instruments which they have at their disposal must then, according to recent organisational and administrative theory, themselves be varied and multiform (Kickert, 1991: 20–21; Godfroij & Nelissen, 1993: 513).

Further bureaucratic differentiation is then a necessary pattern of reaction. That differentiation affects the organisational structure of public administration, the design of and involvement with policy processes, the selection of steering instruments and the results of policy making (see Frissen, 1991).

Theoretical insights in respect of the networked nature of the actors involved in policy processes and in respect of the need for consensus and coalition for the effective conduct of policy are translated into practical terms in the concrete processes of policy making. (See Bekkers, 1993: 59ff.; De Bruijn & Ten Heuvelhof, 1991: 23 ff.) Coalition-forming, institutionalising of networks, designing and arranging structures, are all expressions of this neo-corporatist tendency in the conduct of policy. In contrast to classical corporatism, however, the multiformity of actors is greater and there is a greater incidence of variation and incidental participation. In this connection, the level of policy steering will also vary according to the nature of the policy domain. Complexity, scale and the available technologies vary with each policy domain. Adequate intervention-levels will also vary accordingly. The result is an extremely multiform system of steering-relationships.

Differentiation is an intrinsic characteristic of modernisation processes. Hence the described patterns of change in steering and organisation are seen as a further refinement of bureaucracy. From this point of view, there is no fundamental change in the relationship between bureaucracy and society, but there are certainly far-reaching adjustments. The question then arises whether ongoing differentiation might not at a particular moment herald a truly radical transformation, or even whether, from another point of view, the process is already under way.

2.9 REVERSING BUREAUCRACY

In the forms of change, experiment and innovation which we have sketched, there is an observable transformation process which, in a number of respects, affects the characteristically bureaucratic arrangement of steering processes in our type of society. In the case of public administration, that transformation process involves organisational structure and policy-making processes.

From Pyramid to Archipelago

The restructuring of government organisations shows similarities to restructuring in other organisations where a wide range of writers have observed a tendency to 'bypass the hierarchy' (Peters, 1992; Morgan, 1993). In the empirical reality of organisations and organisation we see a growing development towards smaller, 'flat', autonomous organisational units with only a few links to the wider organisation of the concern or parent company. Management guru Tom Peters has provided a descriptive and prescriptive formulation of this trend (Peters, 1992). On the one hand he observes that large organisations are too clumsy and slow to respond flexibly to rapidly changing markets, which demand a constantly changing range of products and services; on the other hand, he observes smaller organisations operating successfully by focusing particularly on the provision of services, generally with a project-oriented approach. In his opinion, non-hierarchical organisations enjoy an advantage because they can react more rapidly to turbulence and variety. (See also Morgan, 1993)

The best example, according to Peters (1992), is the professional organisation. Because professionals can not be steered hierarchically; because their effectiveness demands autonomy; and because depth of knowledge both for production and as added value is so important, the professional organisation is the organisation of the future. In other words, all organisations will end up like universities and hospitals.

However, if complexity, variety and turbulence are the environmental characteristics which necessitate non-hierarchical organisation, then, *a fortiori*, this must also apply to public administration. After all, it is pre-eminently public administration which operates in this type of environment. Thus many of the changes in forms of steering and organisational structure can be seen as attempts to create smaller-scale, autonomous government organisations and more horizontally-oriented interactive patterns of policy-making. (See also In 't Veld, Noorman & Van der Zwan, 1990.)

The outcome of such developments is a public administration with many more of the characteristics of an archipelago than a pyramid. Numerous, scattered administrative units, varying widely in their functions and tasks are then each responsible for discrete elements of the political domain. Unit autonomy is coupled with the self-steering capacities of society. In a philosophical sense, this development is related to libertarianism. As a basis for this conception of organisation De Geus makes the following claims:

– The general level of education of the population is still rising, making hierarchical organisation a less obvious choice.

- Unpredictability and complexity require 'looser and more flexible' organisation patterns.
- In a libertarian organisation, the power of the centre is limited which prevents 'oligarchy-formation'.
- Libertarian organisation is more democratic and strengthens the learning capacity of organisations.
- Libertarian organisations produce fewer 'authoritarian ideas and personalities'.
- Libertarian organisations are more humane (De Geus, 1989: 225–227).

Strategic Policy Implementation

Linked to the foregoing discussion of lessening hierarchy, is the 'reversal' of the policy-making cycle. In organisations, a distinction is frequently made between strategic, tactical and operational policy making. This distinction is generally linked to a hierarchical pyramid. The hierarchical top of an organisation is responsible for formulating strategy – the general direction; middle management formulates tactical policy; and finally the shop-floor is responsible for operational policy. In the traditional view of policy making this hierarchical perception is recognisable in the distinction between policy development, policy decisions and policy implementation. Here too the hierarchy is decisive in the sense that policy is determined centrally after consultation and advice from 'staff sections' , and then implemented locally. Finally, the hierarchical conception can be detected in the widely accepted cognitive structure of the completed policy, involving a reflective hierarchy of ends and means – the 'goal-tree' and policy theory. (See Hoppe & Edwards, 1985.) However, with increased complexity and turbulence, where organisations become more professional and varied, hierarchy is less satisfactory as a design principle. In these circumstances the constituent principle underlying policy-formulation changes – it is no longer hierarchical. This has implications for the relationship between types of policy and levels of organisation, the relationship between policy development, decisions and implementation, and the relationship between goals and means.

In professional organisations within complex environments, strategy-formulation in the organisation 'sinks down'. If we consider the prototypes of the professional organisation – hospital and university –, strategic decisions, certainly in turbulent circumstances, are taken at the level of the 'shop floor'. That level, in such organisations, is notably also the most highly qualified (the professor and the medical specialist). The highest level in such hierarchies (Board of Governors, Management Board) has mainly facilitating and tactical responsibilities. Many parts of public administration as well, bearing in mind their task environment and qualification levels, can be characterised as professional organisations in turbulent circumstances. Here too one can assume a similar relationship between strategy formulation and tactical and facilitating policy-making.

The same could be argued for the relationship between the development, definition and implementation of policy. In complex and turbulent circumstances, the relationship changes to such an extent that development and implementation become much more closely linked. The feedback from implementation to development intensifies so that implementation can be adapted as quickly as possible and development can be kept as realistic as possible. In public administration, too, we see this need for intensive feedback between policy development and implementation brought about by greater complexity and 'knowledge intensity' in implementation. In this connection, it is interesting to note the plans to implement policy through independent agencies which we have already considered in some detail. In the light of the foregoing, the use of independent agencies seems the obvious way forward for the entire policy-making cycle rather than merely in policy implementation.

Finally, there is the relationship between goals and means in policy-making. It can be argued that in complex and turbulent circumstances meaningful goals can only be established from within these circumstances and by actors who are closely involved in them. Involvement, interest and expertise are thus directly connected. Meaningful goal definition can not occur exogenously or hierarchically; only the governance of means and conditions. In such cases one often hears the term 'meta-steering': indirect steering of structural and procedural conditions under which direct and meaningful policy formation occurs (see Snellen, 1987).

In all these aspects of policy-making we can see a reversal, parallel to that which we observed in hierarchies. Policy-making, too, becomes less or even non-hierarchical and takes on a more circular or network character (see also Hummel, 1990). Intrinsic and strategic primacy is decentralised in the implementation of policy. Huigen rightly refers to 'strategic policy implementation' (Huigen, 1994: 145 ff.). He formulates the core of the argument as follows: 'Implementation processes that can be characterised as political processes focusing on issues characterised by normative ambiguity and factual uncertainty, in which a multitude of policy actors must co-produce policy outcomes.' (Huigen, 1994: 150)

Self-governance: Autonomy and Contracts

The emergence of an archipelago of administrative and social units, and the reversal of the policy-making cycle in a situation of decentralised, strategic policy implementation are closely linked to empirical reality and the conceptual acknowledgement of self-governance and self-organisation in social domains. From this there arises a remarkable reduction in the top-down character of public and political decision-making. In its place there comes a recognition of the network character of policy domains and a corresponding arrangement of the policy-making process by public and private actors.

In 't Veld has linked the increased interest in self-governance, self-regulation and self-organisation – more popular with organisational theorists than in administrative theory (see Peters, 1992; Morgan, 1993) – with three relevant developments and debates, all of which see the effectiveness of central, top-down government interventions as problematic:

- the discussion about state tasks in our kind of society;
- the discussion about new forms and conceptions of governance;
- the debate on the organisation of the public domain (In 't Veld, 1993b: 60–61).

In 't Veld rightly observes that in each of these debates, self-steering is defined differently. However, he rejects the possibility that self-steering implies a reduction of steering relationships. Rather, the number of interdependencies tends to increase. It is much more a case of a reorientation in those interdependencies from rule-led behaviour to forms of contractual relationship. The most important distinction is naturally the hierarchical characteristic versus the horizontal. Contracts are made between equal parties.

Characteristic of self-steering, in my opinion, is that instead of hierarchical regulation a range of contractual relations between private and public parties arises. Part of these contractual relations also constitute agreements about the division of responsibility in order to obviate that well-known feature of hierarchical relationships, 'passing the buck'. (See also In 't Veld, 1993b: 66–67).

Reversals: Postmodernisation?

The rationalisation and differentiation of bureaucracy are typical modernising strategies. They belong to a rationale of improvement which is typical of the process of modernisation. Rationalisation and differentiation are linear in nature: they suggest progress and continual improvement in the design and functioning of organisations. Rationalisation and differentiation also seek increased effectiveness and efficiency and ultimately perfect control. They fit, in other words, into the 'grand narrative' of politics and bureaucracy which, hierarchically ordered and democratically legitimised, lead the process of social development. That narrative aims to provide consistency and coherence in the middle of chaotic and unpredictable developments. The narrative acquires that consistency and coherence from a perspective of democracy – the best political system – and bureaucratic control – the best administrative system – while at the same time honouring the free market as the best economic system.

However, that grand narrative is becoming increasingly problematical, among other things because of the reversals sketched above: from pyramid to archipelago, from a hierarchical to a circular policy-making cycle, from central steering to self-governance. In that context, the politico-administrative system is quite inadequate precisely because it only knows a pyramid framework, its policy-forming is hierarchical and it regards self-governance as a threat. Political primacy – as the constituent principle of the existing politico-administrative system – is the most important explanation.

Earlier we saw that the primacy of politics is also subject to change – it is being restricted, highlighted and reoriented. This process will continue if we take seriously the structural and functional developments occurring especially in professional organisations. Those developments affect public administration in particular because it is one of the most complex, varied and professional organisations known to Western society.

Whether this will also lead to the postmodernising of public administration is still unclear. Ongoing developments have not yet crystallised out; some of them, like technology, have not yet been described; postmodernisation itself has not yet been treated in depth. In the following chapters, I shall explore these issues further.

3 Technology

3.1 INFORMATISATION: AN OUTLINE

Informatisation is a complex of organisational developments arising from the introduction of information and communication technology (ICT) into organisations and the relations between them. This is the definition used by the Tilburg/Rotterdam research programme 'Informatisation in Public Administration'[1] and one which I shall adopt as my starting point adding a few nuances of my own.

Definition

Informatisation, as a generalised concept, indicates the following processes:

a. the introduction of ICT to shape important areas of information-provision with the help of computerised information systems;
b. the introduction of specific expertise in the field of information and communication technology in the form of officers, departments and organisations with powers and responsibilities in that field;
c. the (re)design of internal and external information-flows and relationships in the provision of data for administration;
d. the development of information management as a differentiated branch of management within and between organisations;
e. modifications in the internal and external organisational structure and in the working procedures of organisations where ICT has been introduced (Van de Donk & Frissen, 1994: 44; see also Zuurmond *et al.*, 1994: 17.)

This definition makes it clear that ICT forms the technical core of a wider complex of organisational phenomena. I am obviously referring to informatisation as it affects actual organisations in society and I shall confine my discussion to that area. (This is not, however, to suggest that only the world of

---

[1] The programme began in 1986 and its findings have been published in detail in Zuurmond, Huigen, Frissen, Snellen & Tops, 1994.

organisations is touched by informatisation.) In terms of its constituent parts, the above definition has remained unchanged for years; but within those parts there has been considerable expansion and refinement which we must first consider briefly.

a. Informatisation is a complex of phenomena surrounding technological development. Those technological developments relate to a process of information-provision which is shaped with the aid of computerised information systems. The original definition of 'information technology' (see Frissen 1989: 14), has now been modified to include communication technology. That conforms better to technological developments in which the communicative aspects of information-provision play an increasingly important role: networks, multimedia techniques, telecommunication, virtual reality etc.
b. This element reflects the great importance of professionals and expertise in the process of informatisation. Organisations are being opened up to new subcultures and new strategic dependencies; this last by the fact that much of the expertise in this domain is brought in from outside the organisation. The definition has also been altered to incorporate ICT and to include 'departments and organisations' as well as persons and functionaries. This reflects the ongoing (social) institutionalisation of the information-domain.
c. Designing, redesigning and making information patterns and relations more explicit are important aspects of the process of informatising organisations. Flows and connections in the provision of data often have to be made visible. Formalising previously informal patterns of data-provision is often necessary because information systems require it. The design and redesign of data-flows and relationships are directed to the provision of administrative information either for steering the primary processes in an organisation, or for governance itself as a primary process. A new element in the definition is the reference to both internal and external data-flows and data-relations, which highlights its inter-organisational and inter-administrative dimension.
d. The emergence of information management is the next important aspect of informatisation. Just as other branches of management (personnel, finance, organisation etc.) are differentiated, the same is happening in the domain of information-provision. Decision-making and the steering of information-provision become a separate domain of policy-making. Again, there is a new emphasis on the inter-organisational and inter-administrative dimension of information-policy.
e. Finally, informatisation relates to changes and adjustments in the structure and working-procedures of organisations into which the technology is introduced. This involves both planned and unplanned adjustments since many changes are unintentional and unexpected - an unavoidable

consequence of ICT. A new element here is the addition of working-procedures to organisational structures since neither can be neglected. There is also the reference to ICT and to internal and external structures and procedures which highlights the inter-organisational and inter-administrative dimension of informatisation.

De Jong (1994) rightly points out that there are some elements missing from the definition. Culture, for instance, is missing. Unless culture is to enjoy a different theoretical status from the other elements this is somewhat surprising, bearing in mind the extensive attention which I myself have given to cultural factors in informatisation (see *inter alia* Frissen, 1989). De Jong also notes the absence of any reference to 'control' and 'people' (De Jong, 1994: 27).

However, because definitions are in a sense always arbitrary, I shall retain the original definition. More detail will be added where appropriate in the description of relevant new developments and the outline of the characteristics of informatisation which follows.

3.2 INFORMATISATION: DEVELOPMENTS

Many speculative discussions of technological development tell us more about the writers than about their subject. For example, promising trends in Artificial Intelligence have been predicted for decades, without any of this promise being fulfilled. Yet, research budgets remain huge because self-interest, technology and science are so closely intertwined. Nevertheless, certain forecasts may be made which indicate the general direction of change without being too precise about the technological choices. So with due caution and restraint, I shall predict that developments are likely to be spectacular in their significance and revolutionary in their effects. (See Frissen, 1993c: 120–121)

The following trends may be observed: increased capacity, improved connections, integration and connectivity, and virtual reality. There is an important connection between them, of great significance for organisations, which I shall explore more thoroughly in §3.7.

Capacity

In the first place there is an inevitable trend towards ever smaller, yet more powerful systems. Largely because of miniaturisation, modern PCs now have the processing power of old mainframes. While costs have fallen, capacity has increased out of all proportion.

This factor is responsible for the fact that the microprocessors in a five-year-old's toy today are millions of times more powerful yet far less expensive than the first electronic digital computer, ENIAC, which crunched its first numbers in 1946. (Rheingold, 1991: 71)

Or as Negroponte neatly put it : 'You can assume that what you wear on your wrist tomorrow will be what is on your desk today, and what filled an entire room yesterday' (Negroponte, 1995: 140)

Individual users will thus become increasingly well-equipped while at the same time the need for large centralised systems with correspondingly large organisations will decline dramatically. Technologically, large bureaucratic organisations will become less and less necessary. The future belongs to small, intelligent and flexible units (see also Peters, 1992). 'Small-scale' has become an important technological option.

Connections

In conjunction with this development, communications technology has become increasingly important. Infrastructures for data communication in its broadest sense are being given priority if only because of the strategic necessity to eliminate the classic obstacles of time and space. At the same time, there is an explosive growth in network technology. Electronic connections within and between organisations are expanding rapidly and, in a number of respects, have become more important than the physical organisations and departments which they connect. In effect, social and administrative networks can, as it were, be represented by electronic networks. The enormous success of global networks like the Internet is evidence of that. It is interesting that the Internet grows spontaneously without either governance or controls, yet has become a standardised facility for international communication. All kinds of applications and reference systems seem to come into being spontaneously (see Mieras, 1994a).

Integration and connectivity

Another important development is the integration of systems, media, and technology. Intelligent links between data, images, sound and other sensory experiences make new technological applications possible. Not only does the 'reality level' of the technology become greater, as in virtual reality, but its interactive use becomes possible. In our thinking about social, economic and cultural relations, the assumption of a dichotomy between humans and machines will increasingly have to give way to a conception of man–machine units as links in the countless chains, great and small, of existence.

The integration of systems, media and technologies will only accelerate the existing practice of 'coupling'. After all, computers are 'designed to couple'. A dividing-line between being able to do more and wanting to know more hardly exists. As the electronic traces which humans leave behind them become ever easier to follow, all kinds of profiles will be constructed and those profiles will be used to serve both noble ends (client orientation) and less attractive goals (control and discipline). And we shall have to accept both as inevitable.

Virtual Reality

Possibly the most revolutionary technological trend is that towards virtual reality. By integrating media and systems, an apparent three-dimensional reality can be created. As Sherman and Judkins neatly put it: 'Virtual reality allows you to explore a computer generated world by actually being in it' (Sherman & Judkins, 1992: 19; see also Mieras, 1994b). It is not just a matter of simulating existing reality, as in flight simulators, but also creating new realities as in entertainment applications. An important consequence is that virtuality will become an increasingly relevant part of our lives and organisations. Technologically-generated reality will become a realistic actuality with real consequences.

All these developments in ICT make it obvious that informatisation has not yet crystallised into any clear pattern of services, products, structures, processes and procedures. So we must look more closely at its characteristics.

3.3 INFORMATISATION: CHARACTERISTICS

In their extensive and important survey of the literature on informatisation and democracy, Van de Donk and Tops (1992) propose the following six characteristics of informatisation, or information and communications technology (ICT):

1. ICT dramatically expands the amount of available information. This is not just quantitative; it also involves improved access to the information.
2. The speed with which information is processed, collected, distributed and selected also greatly increases. Limiting factors of time and space apply less and less. Immediacy and proximity become technologically more realistic.
3. ICT enables receivers of information to determine more easily at what time they wish to receive what information. Their control capabilities increase because they have at their disposal more media and better selection tools. Consumer sovereignty becomes greater.
4. Similarly, the senders of information can be much more selective. They have more and improved opportunities for 'information-targeting': in-

formation which is relevant and of interest to specific target-groups. 'Broadcasting' gives way to 'narrowcasting'.
5. ICT promotes tendencies to decentralisation because individual systems are becoming ever more powerful. But at the same time, management and ownership of, for instance, the infrastructure are becoming more concentrated. Van de Donk and Tops speak of a form of decentralisation within a centralised framework.
6. ICT facilitates more interaction between sender and receiver. That distinguishes ICT from the one-sidedness of traditional media. The receiver takes an active part in the game of information exchange and communication (Van de Donk & Tops, 1992: 38–39).

Bekkers (1993) has produced a different survey of ICT's characteristics which he approaches from the perspective of informatisation as a control mechanism. From this perspective, informatisation is aimed at controlling social processes, especially those of differentiation (Bekkers, 1993: 128; see also Beniger, 1986; Frissen, 1989). It is a socio-cultural construction with functional rationality at its core and discipline as its chief goal. Here there is a clear *Wahlverwandtschaft* between informatisation and bureaucracy (Bekkers, 1993: 124–128; Frissen, 1991: 8). Bekkers lists the following characteristics:

Calculation

Because of the hugely increased capacity for processing, collecting and storing data, computers can now calculate and recalculate. Through ICT, complex mathematical models can be used, for instance, to determine the outcome of transactions. Calculation can also be applied to the further rationalisation of the decision-making process. Not only can policy alternatives be worked through, they can even be created.

Control and discipline

Informatisation leads in the first place to a structuring of procedures and regulations. One result of this is greater standardisation and formalisation in organisations, with routine as the effect and objective. Organisations thereby become more stable and predictable while controls and discipline increase accordingly. Informatisation also expands the feedback potential in organisations. All information systems 'quasi-spontaneously' generate information about their own achievements and monitor them (see also Zuboff, 1988: 9 ff). That too promotes control and discipline.

Transparency

Informatisation increases the transparency of structures and processes. Calculation and checking make prediction and control possible and so lead to transparency. It occurs directly where data provision is the primary process and indirectly where it is a secondary or supporting process. Monitoring is intrinsic to ICT.

At the same time, control implies greater transparency. Numbers provide more insight into the wheeling and dealing of organisations and also expand the potential for information and control.

Communication

Finally, communication is an important quality of ICT. Constraints of time and space, which were largely responsible for the creation of organisational structures in the first place, are becoming less significant. Information is now available 'on line' and it is possible to communicate with everyone on a network, irrespective of distance (see also §3.2).

How these characteristics should be evaluated is a subject of debate. That debate – 'the technological debate' – has not always been very illuminating. Nevertheless, I shall briefly discuss it because the positions which have been taken up raise the important question whether the characteristics of ICT are inherent or situationally determined. Are they context-dependent or not? This is important because on the answer hinges the more important question whether public administration is being left any significant freedom of choice.

3.4 THE TECHNOLOGICAL DEBATE: POSITIONS

Is technology an autonomous factor in society or can it only be understood in the light of the objectives set at the time that the technology was developed and applied? To that question the so-called technology debate is seeking an answer. Zuboff (1988) articulates the issue by speaking of two dimensions of technological change, the intrinsic and contingent. Do technological changes, by virtue of their intrinsic qualities, lead to specific social changes? Or are those changes contingent on the existing context of choices within which the technology is applied? In her own words:

> To fully grasp the way in which a major new technology can change the world (...) it is necessary to consider both the manner in which it creates intrinsically new qualities of experience and the way in which new possibilities are engaged by the often conflicting demands of social, political and economic interests in order to produce a 'choice'. To concentrate only on intrinsic change and the texture of an emergent mentality is to ignore the real weight of history and the diversity of interests that pervade collective behavior. However, to narrow all discussion of tech-

nological change to the play of these interests overlooks the essential power of technology to reorder the rules of the games and thus our experience as players (Zuboff, 1988: 389).

In this debate, three positions can be identified. I shall discuss them in relation to informatisation (see Van de Donk & Frissen, 1994: 45–49).

Determinism

Informatisation is a strategy or strategic development rooted in the characteristics of the technology – in this case ICT – which has autonomous consequences for social reality. From this perspective, ICT is not neutral; its effects are not primarily dependent on the prevailing goals but deterministic. It has, as it were, its own goals. It gives autonomous form to organisational structures and processes and imposes compelling demands on the context in which it is applied.

In earlier publications (*inter alia* Frissen, 1989), I argued that informatisation is a greedy process in which technology based on functional rationality expands autonomously within the framework of a control-oriented technological culture. ICT, because of its standardising and formalising characteristics, is itself inherently bureaucratic. Technical systems determine social systems. The objections to this approach apply particularly to current conceptions of social and technological dynamics and of technology. Both are too simplistic. They reduce social and technological dynamics to a uniform causal process of social change determined by technological development. This perception of technology is too abstract and one-sided. There is too little eye for ambiguity and variety and the wide differences between ICT applications are ignored.

Voluntarism

In this approach, ICT is a neutral and malleable instrument which can be used to achieve specific results according to objectives laid down by users, organisations or developers. The consequences of development and use are always the effects (intended or unintended) of the ambitions of those who decide to develop and use ICT in concrete instances. Informatisation, therefore, is not a strategic process with intrinsic characteristics or objectives. The technology is neutral; it has no 'intentions'. ICT is a collection of tools, which enables individuals and organisations to realise certain goals.

From this approach, the results of informatisation are always explained by the dynamic of change in the system and in society. Informatisation then becomes, for example, a process which can only be understood through existing power relationships (see *e.g.* Danziger, Dutton, Kling & Kraemer, 1982). The most important weaknesses of this approach are its technological and cultural naïvety. Voluntarism is technologically naïve because it misunderstands the social dynamic of technology: every observed effect is traced back to non-technological intentions or causes. Voluntarism is culturally naïve because, in the first place, it does not recognise that technologies are cultural artefacts and secondly, it fails to see to what extent our culture is a technological culture.

Constructivism

This perspective on technological development is more historical and more dynamic than the previous two. The relationship between informatisation and social change is not perceived as causal (setting and achieving objectives with the help of technology), nor as deterministic (the effect of the intrinsic characteristics of technology). From this point of view, both technological developments and social change are the results of interaction between actors and their normative patterns of behaviour. (See *e.g.* Bijker & Law, 1992) Actors make choices, but do so within contexts which condition those choices and within a framework of repertoires which the actors have created historically. Technological developments and power relations within social systems are therefore both cause and effect of these interactions.

From this viewpoint, it is not so much the consequences or effects which are central, but rather the process; a process which combines stability and dynamism, determinism and choice, objectives and preconditions. Technical and social systems interact continuously and through that interaction they influence and condition each other. Its results are, in principle, provisional and more specifically tied to time and place (Van der Meer & Roodink, 1991).

The objections to this approach are that it tends towards a decisionist perception of reality and that it neglects technological constraints. Constructivism therefore looks like a sophisticated version of voluntarism. Choices, interaction and behaviour receive greater emphasis than technological, social and economic conditions.

3.5　ADVANCED CONSTRUCTIVISM OR DETERMINISM?

Of course, it is not essential to adopt a particular position in the technology debate: the verdict can be left to the reader. In any case, the manner in which the debating-positions are usually presented seems inevitably to lead to a victory for constructivism. In a dialectical sense, this is a sensible synthesis of the other two positions (see Van der Meer & Boer, 1994).

Nevertheless, a more precise definition of my own position will be helpful in the light of the theoretical discussions of politics and governance which follow.

- It is necessary to bear in mind actual technological developments. Those developments reveal trends (see §3.2) which are not the outcome of some grand design or blueprint, but are the result of history, constraints and decisions. Relevant trends in technological development build on the fundamental characteristics of technology. At the same time, a selection is made from many possible technological developments, which can be explained by the social context in which the technology is applied, such as political and social power relations, economic circumstances and contingencies. Technological development, then, forms a totality of 'best practices' which – during sudden change – function as cause and 'attractor' of new institutional arrangements. (See Perez, 1983)
- Informatisation is a development which admittedly can not be separated from interactions involving its design and application, but which as a macro-development can not simply be traced back to the choices of groups and individuals. As macro-development, informatisation forms, as it were, the sum total of decisions and other developments, and thereby becomes a determinant of new social developments. The constructivist standpoint is in fact voluntaristic since it refuses to see the result of all constructivist interactions as deterministic phenomena.
- Informatisation is not merely one of a number of developments which help to explain social dynamic. We are dealing here with a development of revolutionary importance because informatisation:
-
  - replaces cognitive, not physical, human capacities;
  - questions the anthropocentric world view;
  - does not merely add information to specific processes, but always generates new information about those processes (Zuboff, 1988:9)

Consequently, we have a technology which is both intellectual and reflexive.

- We live in a technological culture which determines our behaviour. In this connection, Poster (1990) talks about 'the mode of information', by analogy with Marx's mode of production. In our culture, technological development has made information and communication the dominant constituents of economic, social and political relationships. Information and communication have become an autonomous discourse, free of their subjects. Indeed, they create new subjects. The relationship between man and machine is thereby changed dramatically.

- We can therefore agree with Coolen (1992) that through technology humans have objectified the struggle to become an autonomous subject. Information technology is an extreme form of this in which the 'self-reflection of rational reflection' achieves objectivisation (Coolen, 1992: 269). At the same time information technology lessens human subjectivity (see also Poster, 1990:7). The nature of technological culture is such that it determines human thought and behaviour because thought and action outside that culture is no longer possible. In a positive sense we can speak of symbiosis between man and machine; in a negative sense, of 'epistemic enslavement' (Van den Hoven, 1994) at the level of society as a whole.

In the foregoing I have sketched some of the principles on which my discussion of technological development is based. I shall now consider more specifically a number of relevant technological developments.

3.6 NETWORKS, CONNECTIVITY AND VIRTUAL REALITY

Informatisation is a widely inclusive process of social and organisational developments surrounding the design and use of information and communication technology (ICT). ICT is in turn itself a comprehensive complex of technological systems and applications. Information systems are often spoken of as consisting not only of hardware and software for collecting, processing, modifying and distributing information, but also more widely as the people and procedures engaged in collecting, processing, modifying and distributing information with or without the help of advanced technology (Brussaard, 1992: 173). ICT can thus include a laptop computer, a mainframe, an organisation's network, an interactive television, a videophone, a fax, a glass-fibre cable, communications standards, protocols, methodologies for system development, a Nintendo 'game-boy', the electronic super-highway and so on.

At the heart of this is the application of technology to information-provision and communication. And this has undergone such dramatic development during the past decades that one now talks of the 'information society', of comprehensive transformation, and of radical modernisation. Such phrases, whatever they may actually mean, reflect the urgency and relevance of ICT for social relationships and processes.

I shall now turn to some relevant applications which illustrate the developments which I sketched earlier (§3.2). They are networks, computer coupling and virtual reality.

Networks

Networks are technical facilities and protocols which make data communication possible. Networks allow 'the direct exchange of data between computers by electronic means' (Perlee, 1993: 62). Data consists not only of characters but also of speech and sound, graphics, film and video. In order for data exchange to take place, certain conditions have to be satisfied (see Perlee, 1993: 62–63):

> *Connections*: between the computers there must be a connection either in the form of cables or radio waves (satellite, transmitters, mobile phones).
> *Transport*: there must be agreement on the manner in which data will be transmitted. This involves addressing, routing, establishing and terminating contact, error detection and correction.
> *Telematics*: there must be agreement on the generic applications used on the network, like electronic mail, telnet, file transfer and management.
> *Contents and structure:* although strictly speaking not part of the network, there must also be agreement on the content and structure of the data to be transmitted. This requires standards for data definition, document structure, the ordering of data elements etc.

Networks are both intra- and inter-organisational. They function at the lowest organisational level – interdepartmental networks – as well as on a global scale – the Internet which connects millions of individuals and organisations. Within or between organisations there is usually a central or common configuration of connections and management to which only authorised persons have access.

In world-wide networks like the Internet, there is little or no central control. It is interesting that the Internet has military origins and that the lack of central control over the network was a deliberate military and strategic decision: communications could then never be disrupted or disconnected. There lies the success of the Internet. It offers an infrastructure for communication along which information can be transported by innumerable and flexible channels. In Rheingold's words:

> This invention of distributed conversation that flows around obstacles – a grassroots adaptation of a technology originally designed as a doomsday weapon – might turn out to be as important in the long run as the hardware and software inventions that made it possible. (Rheingold, 1994:8)

Connectivity and Computer Coupling

A second important application or facility of ICT, which in a number of respects is made possible by combining network technologies, is the constant improvement and ease of connectivity between databases. Developments in the field of relational databases and database-management methods and systems have also contributed to this.

Stimuli for connectivity are varied:

- From the point of view of control and verification it is efficient and effective for organisations to be able to compare data from different databases.
- Combining databases can enrich the information which organisations have at their disposal.
- Merging data about entities, especially people, can provide insight into patterns and relationships.
- The identification of patterns and relationships makes it possible for organisations to develop new policies.

In both public administration and business, the coupling of databases has increased considerably in frequency and intensity (see Van Duivenboden, 1994; Mieras, 1995). In public administration it was brought about by the consensus on the need to combat fraud. In the business world, coupling is particularly popular in marketing: marketing bureaus couple data from transaction and payment systems to identify sections of the market for specific products. Banks too use the front-office systems to build up client profiles and develop new combinations of products and markets. And recently there has been the phenomenal success of the Air Miles scheme.

Four types of coupling can be distinguished: (See Van Duivenboden, 1994: 398ff.) The first is 'front-end verification'. Data from various databases about known persons are exchanged. For that purpose, the social security number was developed, and its application has been eased considerably since the Local Authority network became operational. It is now possible to exchange large quantities of information about individual citizens efficiently and effectively.

A second form of coupling is 'computer matching'. Information is exchanged between computers about unknown individuals, with the intention, of course, of making them known. This is done on the basis of a distinguishing feature or some assumption about them. A well-known example is the case where a local social services office suspected that a number of 'unemployed' hairdressers were in fact moonlighting. There seemed to be no way of finding out until a social security officer came up with the bright idea of comparing their own database with that of the public utilities. The assumption was that unemployed hairdressers would have excessively high domestic water bills if they were working from home. This coupling of files identified precisely those hairdressers who were working illegally.

A third form of coupling goes a step further, and is known as 'computer profiling'. This makes it possible to search through databases without having any predefined or specific result in mind. All kinds of data are combined to obtain interesting population profiles. One example of this is the welfare index of the town of Enschede in which all kinds of files with information about social deprivation are coupled with each other. On that basis, a distribution ratio is established and used for the rational allocation of welfare subsidies to the various districts in the town according to the degree of deprivation. At the same time, of course, the people who are suffering deprivation are carefully registered.

The fourth, most far-reaching form of coupling is the permanent integration of different databases, by which new registrations are produced. As relational databases are developed further this form will be of particular interest. We will then no longer speak of physical databases, but of databases which only exist 'virtually' on the screen of the user. Legally, too, this is an intriguing phenomenon.

Virtual Reality

A third application of ICT is Virtual Reality. Virtual Reality (VR) is one of the biggest hypes in ICT, because on the one hand its domain of application seems to have no boundaries and on the other its cultural and social implications are enormous.

The state of the art and future developments and implications are well surveyed in Sherman and Judkins (1992) the title of whose book, 'Glimpses of Heaven, Visions of Hell', neatly reflects their approach. VR has both extremely positive and extremely negative consequences. (See also Negroponte, 1995: 112 ff.)

VR is based on graphical representation, though film or video can also be used. VR always has three elements: 'it is inclusive, it is interactive, and it all happens in real time' (Sherman & Judkins, 1992: 20). In other words, the user participates, can introduce changes and observes the changes immediately. Sherman and Judkins have distinguished three forms of VR:

1. The first form of VR makes use of a helmet with small TV screens and headphones, a glove, a joy-stick and a staff or a 'six-dimensional' mouse.
2. In the second type of VR, video-cameras place and follow the image of the user in a virtual graphical world. The user interacts with virtual objects. The video image can also be converted into a graphical image which is subsequently placed in the virtual world.
3. The third type is a form of three-dimensional modelling, which is viewed through 3-D glasses or projected on to a large, curved or bent screen to obtain the 'inclusion-effect', better known as 'immersion' (Sherman & Judkins, 1992: 20).

Applications of VR can be imagined in widely diverse domains. In the field of entertainment different games are already available, in amusement arcades for instance. But the really interesting possibilities are in the form of interactive TV. VR can be used for a wide variety of simulations and is thus suitable for training and education. Military skills are already being taught by this means. VR also offers facilities for the world of design (industrial, architectural, technical, infra-structural, urban and rural planning). Furthermore, VR will undoubtedly be combined with other ICT applications, for example in the form of interfaces, or with artificial intelligence and other technologies (Sherman & Judkins, 1992: 21–22; Rheingold, 1991: 154 ff.; Negroponte, 1995: 112 ff.).

Every VR system consists of one or more computers, sensors (input devices), display arrangements and software. The central computer produces three-dimensional graphical images, possibly in combination with graphics boards or rendering engines. Users see these images on a display possibly supplemented by sound and touch-feedback by means of the glove or joystick. Users can then interact with the images via various input devices which stimulate the computer's sensors. The sensors observe the body movements of the user and hear his voice and are able to record video images. Whenever the sensor observes a change, the entire computer-system (the 'reality engine') generates a new set of images representing the change. For a system to be 'realistic', the computer must sample its sensors about sixty times a second to show real-time changes, apparently immediately but in fact with an extremely short time-lag. The user sees, hears and perhaps even feels these changes on the display and reacts accordingly, which is again represented graphically (Sherman & Judkins, 1992: 42-43). On the Internet there are all kinds of discussion groups, newsgroups and mailing lists dedicated to VR.

After this brief survey of a number of relevant ICT applications, we now turn to their organisational and cultural meaning.

3.7 ORGANISATIONAL MEANING

Informatisation has far-reaching consequences for organisations and organising, as has been argued by many other people in many different ways. The important question here is whether informatisation is a continuation of the age-old process of modernisation – another chapter in the history of the progress of civilisation – or whether it marks a qualitative break with that history. Because many of the developments are so recent, a definitive answer can not yet be given. Furthermore, the speed of change is such that what is mere speculation today can be actuality or even out of date tomorrow. A simple look at the average work-place of a civil servant or university lecturer reveals a dramatic contrast with a mere five years ago, while one may also reasonably expect that in another five years the software and hardware currently being used will have been entirely changed or replaced. The possibilities will again have increased exponentially.

The consequences of informatisation in general can be categorised as follows: refinement, transparency, horizontalisation and virtualisation.

Refinement

Informatisation merges seamlessly with every effort made to improve organisations and their functioning. Morgan (1986) sums it up with his image of the organisation as a machine. Ever since the age of mechanisation, attempts have been made to design and control organisations like the machines which constitute the production process. The assembly line as the most symbolic expression of mechanised production seems to have become a metaphor for the design of the entire organisation.

Standardisation of activities, specialisation of tasks, formalisation of authority and centralisation of command and power are the characteristics of the bureaucratic ideal-type as well as the normative guidelines for the style of organisation known as Taylorism and Fordism. Automating as many tasks and activities as possible is a basic precondition for mechanisation and achieving benefits of scale through mass production.

Information technology is based on just these characteristics. Standardisation and formalisation, in particular, are the technical preconditions for the successful development and implementation of information-systems. However, I.T. is not merely a tool for mechanising and automating the provision of information as production process like the transaction systems of banks, insurance and government, or as production support and process management. What is really radical about informatisation is that it unites information-provision, governance and automation. Informatisation always provides control and steering information about the processes which it automates (Zuboff, 1988: 387ff.)

In this sense informatisation contributes to the further refinement of the organisation conceived of as a machine and as the embodiment of the pursuit of domination. As early as 1979, Inbar prophesied:

> Because a computerized bureaucracy might come to be viewed by many as psychologically less frustrating than a clerically staffed one, as socially more equitable, as technically more reliable, and because it holds the promise of a partial solution to some of the unresolved problems of representativeness and validity, the notion of an automated bureaucracy may well gain growing support. (Inbar, 1979: 208)

The survey of empirical research by Zuurmond *et al.* (1994) shows that we have already moved a long way down that path.

Transparency

The effect of transparency was described in broad terms in the famous report by Nora and Minc (1978). In their opinion and that of many others, automating the provision of information will lead to greater, perhaps even complete, transparency of social systems.

Snellen (1994: 286–287) explores this in detail. Because organisations now use information systems in many of their activities, these activities are now recorded. This is the consequence of what Zuboff calls the 'informating' aspect of ICT:

> (...) the same technology simultaneously generates information about the underlying productive and administrative processes through which an organization accomplishes its work (Zuboff, 1988:9).

Snellen argues that this feature of informatisation, together with rapid developments in database technology, makes possible the application of all kinds of heuristic techniques to the mass of data which organisations now possess. This includes data about their own production processes, about the administration and steering of these processes, as well as about the environment, clients, markets and other relevant exogenous developments. This process is now known as 'data mining'.

Data coupling, of course, greatly enhances this process. The transparency of organisations and their surroundings increases sharply when diverse databases are interconnected. In this way, for instance, 'computer profiles' of clients and markets are constructed for commercial purposes as well as for purposes of control and management. (Snellen, 1994:9) The monitoring of organisations and social change becomes much easier. Combining databases, networks and multimedia enhances transparency within and between organisations. The integration of tasks and the removal of functional differentiation can also be expected (Snellen, 1994; Scheepers, 1991). Transparency is therefore also a consequence of the network technologies which we considered earlier. Through networking, organisations become transparent in respect of each other and to different environments.

An important aspect of transparency is that organisational identity, in so far as it is based on data, information and knowledge, becomes blurred. Of course, property rights continue to exist and security requires the demarcation of boundaries. But the interconnections are now so powerful, that organisational boundaries become more fluid and increasingly temporary in character. The revolutionary component in the impact of transparency could be that organisations as spatially discrete and 'informationally' delineated systems will become less important. I shall return to this later.

Horizontalisation

One consequence of network connections within and between organisations is that it changes the patterns of inter- and intra-organisational information-provision and communication. In the effective, if typically overcharged language of an American management best-seller, Peters formulates it thus:

> Bits of the changing nature of 'organizational' relationships include (1) demonstrating trust (a willingness to share virtually *everything* with *everybody*, inside and out); (2) creating on-line databases that can be used across functional boundaries (to the extent that old functions will even exist anymore); (3) installing an 'e-mail ethos', where informal communication across remaining levels and functions becomes normal; (4) hooking into on-line databases and electronic bulletin boards external to the firm; and (5) using electronic data interchange (EDI) extensively to routinize and automate transactions with 'outsiders'. (Peters, 1992: 122)

Within organisations, networks lead to easier access to databases for all employees, thereby rendering unnecessary many of the classical forms of vertical task-specialisation. Data distribution along networks will lead to the disappearance of middle management with its obviously intermediate position in the flow of information (Snellen, 1994). In terms of hierarchy, it means a flattening of organisations and decreasing distinction between the execution and supervision of tasks. The consequence is a more horizontal structure. The process is intensified even more because networks encourage communication patterns which are not tied to a hierarchy: the 'e-mail ethos', referred to by Peters, for instance.

Horizontalisation from an inter-organisational perspective is more far-reaching still. Networks between organisations, particularly at the international level, can make the horizontal relationships between parts of one organisation and parts of other organisations, even on a global scale, more important than the traditional links within the vertical hierarchy of a single organisation.

Because all kinds of other developments which are not strictly technology-driven also show signs of horizontalisation (outsourcing, autonomy, 'mean and lean' production, 'core business', contracting etc.), a powerful coalition of technological and organisational trends is moving in the direction of organisational and social networks which are underpinned by electronic networks. As was pointed out earlier, the basic principle of the Internet and the reason for its success, is the lack of central control. In Rheingold's words, it is in essence a 'grass-roots' technology (Rheingold, 1994: 9).

These infrastructures for data communication have little in common with the 'classical' infrastructure. It is precisely because they distribute information and allow communication on line and in real time that they facilitate organisation patterns with a strongly horizontal character. Various commentators have remarked on the rise of 'network organisations', inter-organisational connections which employ inter-organisational systems for communication and information-provision. These network organisations operate in electronic markets (Malone, Yates & Benjamin, 1987; Schmid, 1993):

> Das würde aber bedeuten dass die heute vertikal hochintergrierten Unternehmungen sich in Richtung eines Netwerkes marktmäßig koordinierter Einheiten entwicklen' (Schmid, 1993: 479). [*That means that today's vertically-integrated enterprises will develop in the direction of networked units co-ordinated by the market.*]

Virtualisation

I understand by virtualisation, firstly the direct consequences of applying virtual reality to organisations, and secondly the implications of all those ICT effects for which the term 'virtual reality' is employed metaphorically (changes in time and space, changes in the experience of reality, changes in physical and social relations).

The direct application of VR inside and by organisations means that reality can be sim· lated by integrating media, coupling databases and using advanced graphical interfaces in order to develop new markets and products, formulate new policies and strategies, create future scenarios and so on. Existing realities are already being more realistically imitated.

VR applications can also lead to the creation of new 'realities' which in turn can lead to the breaking down of existing patterns in product-development, marketing and policy-making. But it is particularly in the field of leisure and recreation that the most sensational developments can be expected. One only has to think of the combination of VR, TV and computer games.

Finally, VR will be used to support interaction between organisations and their environment. It will enable clients to become more involved in the design of products. Because the senses can be stimulated more fully, clients will be able to experience and observe a new product more completely, and ensure that the design conforms more fully to their personal preferences (Peters, 1992: 117–118).

The second meaning of virtualisation is metaphorical and refers to the effect of ICT in changing perceptions of reality and physical and social relationships. My starting-point is Sherman and Judkins' dictionary-based definition: 'Something which is real, or has actual existence, but not formally or actually.' (Sherman & Judkins, 1992: 164)

This highly concise definition indicates a possible and apparent reality, which is experienced as such, but has no actual or independent existence beyond the observer. Building on this definition, virtualisation is a complex of developments by which organisations and social and physical relations will become 'virtual'. This is neatly illustrated by John Perry Barlow, songwriter of the Grateful Dead: 'Cyberspace (Virtual Reality) is where most of your money is, most of the time.' (Sherman & Judkins, 1992: 162)

Does this money exist? Most certainly, since serious economic crises can result from problems in Wall Street's financial cyberspace. But our imagination fails us when we try to visualise this cyberspace in images of piles of bank-notes.

Snellen (1994) refers to virtual databases which are constructed from diverse and physically distanced data stores and which only exist fleetingly for the user who creates them.

Another example is computer conferencing by which an interactive and social environment is created but which from the perspective of the participant does not in fact exist in a specific space-time constellation. It is a very effective form of virtual organisation, but 'one that, by its very nature, respects neither territorial boundaries, nor subjective loyalties' (Mowshowitz, 1992: 286).

It is precisely this combination of diverse technological applications such as networking, data-coupling and VR which contributes to the virtualisation of organisation patterns and social relationships. In summary then, virtualisation relates to the following aspects:

- Technology is getting better and better at simulating existing reality by integrating technologies and media.
- Technology is getting better and better at creating new realities by the same means.
- Because man and machine become, as it were, symbiotic entities in a wide range of transactions, technologically simulated or created realities are experienced by individuals, groups and organisations as increasingly realistic.
- Integration of technologies and media makes distance in time and space more relative: on-line connections over long distances are possible; limitations on simultaneity and proximity are reduced.
- The concept of a delineated 'territory' within which transactions and communication must occur becomes less relevant.
- Communication and business transactions can now be designed at any desired level of scale, participation and information-provision.

Virtualisation has important consequences for organisations. In the first place, delimitation of organisations to reflect their physical and bureaucratic identity becomes less important. Relationship patterns with other actors are not only given more weight but become substitute organisations. The network becomes an important, perhaps the most important, work context, both technologically and socially. The common use of databases, applications and communication channels and media becomes more important for defining production and policy processes than the classical, demarcated organisation. Because of technology, information and communication links are less and less constrained by time and space. Organisations thus become more fluid and more flexible. They can be redesigned to satisfy specific needs or specific interests.

In the second place, virtualisation means that simulated and constructed realities will become increasingly important inputs. Whether it is a question of information generated by the coupling of databases or a three-dimensional representation of a product, intelligent and creative systems will enjoy a degree of autonomy in defining the agenda.

The foregoing observations suggest that technological arrangements will take priority over organisational arrangements and that technological constructions of reality will become authentic policy-making realities. Conceptually this requires a new approach to organisation theory; practically it will require the redesign of organisations (Frissen, 1994b: 444).

3.8 CULTURAL MEANING

Analogous to the preceding sketch of the organisational meaning of informatisation in general and some specific applications in particular, is its cultural meaning. In other words, informatisation and ICT not only have implications for organisational arrangements and social relationships, but also, and perhaps especially, have an impact on the patterns of meaning surrounding these arrangements and relationships.

Modernisation and Bureaucratisation

As I have shown elsewhere, informatisation strengthens the bureaucratic culture of organisations (Frissen, 1989). The basis for this was the ideal-typical *'Wahlverwandtschaft'* or affinity between the culture of bureaucracy and the patterns of meaning embedded in information technology. The process of introducing and developing information systems demands standardisation and formalisation. The processes which are the object of informatisation must also be capable of being standardised and formalised. The application of information systems is only possible to the extent that this form of stability has been achieved. These technical requirements, however, always contain a normative connotation. Standardisation and formalisation are not just conditions for effective system development, they are also evaluative categories in respect of organisational processes. On the one hand, they indicate a stability which is desirable and worth striving for, failure to achieve this being seen as evidence of carelessness or lack of control. On the other hand, there is the promise implicit in the technology that at some time in the foreseeable future it will be applicable to non-standardised and informal processes.

In other words, we are dealing with a technological culture which is not only created by the existing context into which information systems are introduced, but which is determined to an important degree by the cultural characteristics of the technology itself.

The technology is greedy; it creates its own application domain and then hunts, without discernible opposition, for new domains. The need for progress always associated with information technology is not under discussion. Cybernetic presuppositions for this metaphor have a long and respectable tradition. They form a part of what Beniger (1986) calls the 'control revolution' of which information, as the most important condition for control, is a crucial element. The unique character of information technology is not just that physical capacities are no longer substituted in order to control physical processes but that it is an intellectual technology (Bell, 1979), which is able to control social processes by codifying theoretical knowledge. In other words: the radical claim of this technology is that it controls social and mental processes. Functional rationality moves up as the dominant pattern of meaning and increasingly replaces the normative framework of substantive rationality (Frissen, 1991: 8–10).

Modernisation and Differentiation

The technological developments sketched in §3.2 have a more far-reaching cultural meaning than the obvious kinship between technology and bureaucracy. After all, ever-increasing differentiation is the outcome of technological developments because of increased capacity, potential for coupling and linking, and virtualisation.

The increased capacity of systems is an intrinsic threat to large-scale, vertically integrated organisations. Entrepreneurship overtakes bureaucracy as the dominant discourse and seems easier to achieve through the increased information and communicative capacities of systems and individuals.

Modernisation's promise to 'demystify the world' also penetrates the organisation now that its machine or mechanical character is softened through the powerful ICT equipment of individuals and organisational units. Peters speaks of future organisations as 'knowledge-based societies' (Peters, 1992: 123) and Zuboff of post-hierarchical relations in the 'informated organization' (Zuboff, 1988: 399).

The further differentiation brought about by ICT and informatisation makes it possible to emancipate the bureaucratic organisation from control and domination by the centre. The technical and organisational principles of the Internet (distribution of information without central controls, a system of autonomous add-ons of participants, applications, supply) are at the same time cultural characteristics of interorganisational patterns of relationship.

Here the norms and values of bureaucracy are replaced by those of the market and of autonomous social systems. Standardisation and formalisation are perhaps still technically required, but more and more they form the conditions for flexibility, creativity and autonomy.

This is no utopia. This cultural change by no means signifies the disappearance of the struggle for mastery and inequality. The international networks are used intelligently by large companies. Mowshowitz (1992) shows that the continual movement of production facilities, sources of income and tax centres by multinational companies is only possible through ICT and that this is also a form of virtual organisation. ICT and warfare are also still powerful allies as was shown during the Gulf war. The difference is that ICT can turn individuals into multinationals and warriors.

The greater autonomy of decentralised units in an organisation due to ICT, occurs within a framework of common and therefore in some respects centralised infrastructure for data-communication. Snellen and Van de Donk (1987) call this dialectical relationship 'decentralisation within a centralised framework'.

Changes in Time and Space

Possibly the most important cultural consequences of informatisation stem from changes in the experience of time and space which these technological developments bring about. Overcoming constraints of time and space has always been the basis for forming organisations. Organisations can be described as time-constrained reproductive processes, physically contained and controlled within a particular territory. It is clear that with the arrival and expansion of ICT, the constraints of both time and space are becoming less relevant. Immediacy and proximity can be achieved permanently through on-line connections throughout the entire world. In this way, ICT contributes to the acceleration of the modernisation process. The acceleration of time and contraction of space through internationalisation and globalisation are the hallmarks of modernisation (see Giddens, 1990: 17 ff.; Harvey, 1989: 201 ff.). Or as Harvey formulates it:

> Modernism entails, after all, the perpetual disruption of temporal and spatial rhythms, and modernism takes as one of its missions the production of new meanings for space and time in a world of ephemerality and fragmentation. (1989: 216)

Modernisation, then, is a process of continuing 'time-space compression'. Time speeds up and space shrinks. With the help of ICT, time-lags can be made so small that in effect real-time communication with the whole world becomes possible. But just because of this speeding up and contraction, there is, in a cultural sense, a '(...) trend to privilege the spatialisation of time

(Being) over the annihilation of space by time (Becoming)' (Harvey, 1989: 273). Progress, so characteristic of modern Enlightenment thought, becomes less important. The local, the particular, the specific can thereby be constantly and repeatedly realised without space-time hindrances. That is one of the central characteristics of fragmented post-modernism (see chapter 7). In the seminal work of Castells this is called 'the space of flows' and timeless time (Castells, 1996–98).

For organisations, it means that size loses importance, at least size associated with vertical integration within a hierarchical entity. Through technological capacity, the scale of organisation can be small at the same time as the scale of operations continues to expand. In other words, the scale of operations can be chosen according to the temporary network-connections required (Peters, 1992:122). At the same time, fluidity and flexibility increase. The scale of organisations and their scope change more quickly and more frequently.

As a result, time–space coherence is steadily losing its importance as is physical organisation as the focal point of loyalty. Organisation cultures, often represented as beacons of unity in fragmented organisational reality (e.g. Deal & Kennedy, 1982: 193), are becoming obsolete at a time when loyalty is focusing increasingly on networks and horizontal connections.

The blurring of boundaries between organisations caused by ICT thereby gains a cultural component. Virtualisation also affects the normative patterns of organisational realities.

Lessening Subjectivity

One of the most fascinating cultural consequences of informatisation and ICT is the changing image of humans and machines. Turkle expresses this fascination as follows:

> Most considerations of the computer concentrate on the 'instrumental computer', on what work the computer will do. But my focus here is on something different, on the 'subjective computer'. This is the machine as it enters into social life and psychological development, the computer as it affects the way that we think, especially the way we think about ourselves. I believe that what fascinates me is the unstated question that lies behind much of our preoccupation with the computer's capabilities. That question is not what will the computer be like in the future, but instead, what will *we* be like? What kind of people are we becoming? (Turkle, 1984: 13)

The computer differs from other machines because it is (already) seen as a living creature with subjectivity. This viewpoint is often metaphorical, sometimes empirical, as in the field of so-called 'strong' artificial intelligence (AI). A combination of ICT, AI and biology and biotechnology ought then to make artificial life possible (Levy, 1992; Kelly, 1994). The opponents of this idea of a living machine are as hostile as its devoted supporters are enthusiastic (Roszak, 1986).

But apart from the rhetoric and speculation, there is the cultural signification of the man–machine symbiosis which, more and more, is the hallmark of social and particularly organisational relationships. Increasingly, modern organisations can only be understood as information processing arrangements within which humans play a role, but no longer necessarily a dominant one. The 'knowledge worker' (Lenk, 1994) can not be envisaged outside ICT. Although there is a hint of historical continuity, there is also a historical break:

> Although profound and nearly invisible processes of economic subsumption have never ceased effecting the convergence of the biological and mechanical spheres of existence, what is changing today (...) is how the classical processes of the mechanization of life are giving way to a new and unprecedented *vitalization of the machine*. This development in no way signals a monolithic historical shift but rather a proliferation at a variety of levels of new virtual pathways and historical counter-movements which have the potential to be used or activated in diverse and opposing ways. Though many of these tendencies are either already stabilized in familiar social arrangements or deeply embedded in the inexorable movements of contemporary regimes of power and production, there are others that remain volatile, paradigm-resisting forces, full of unknown and unforeseeable capacities for cognitive and cultural transformation. (Crary & Kwinter, 1992: 14)

The historical break is particularly visible in the following cultural consequences of the symbiosis between man and machine in the information era.

Decentring the subject
In the electronic era, the human presence is scattered and fragmented in network files. Through this diffusion and fragmentation, the subject as a meaningful entity is decentred and multiplied (Poster, 1990: 7). Anonymity and a choice from multiple personalities is typical in international networks (Poster, 1990: 119; Stone, 1992: 611). Here too we see striking parallels with postmodernist thinking. (See too Baudrillard, 1985: 127–128.)

Anthropocentrism
The growing capacities and quality of ICT are increasingly leading many organisations to replace human actions by computers. An everyday example of this, but no less culturally significant in its impact, are computer-generated decisions in public administration (Snellen, 1993). Human decision-makers are no longer necessary because the machine can do it better since it is faster and more objective. Expansion into other areas is naturally not excluded. The image of man as the centre of the world is thereby diminished.

Epistemic enslavement
The increasingly strong symbiosis of man and machine in social contexts leads, according to Van den Hoven, to 'epistemic enslavement'. The machine inevitably determines and limits human freedom of decision. This is the case if system users and the systems themselves are placed in an institutional arrangement of 'narrow embeddedness'. This occurs when there is no opportunity for ample consideration and judgement (Van den Hoven, 1994: 358–360).

In later chapters, I shall explore the cultural impact of ICT and informatisation more fully.

3.9 DIGITAL AMBIGUITIES I

The idea that technological developments follow a straight, prefigured path of progress is completely wrong. On the Internet, AIDS sufferers can exchange news (Mak, 1994), Amnesty International can publicise the names of missing political activists, child pornography is distributed, the extreme right plans bombings, numerous corporations have web-sites, while commercial use in general increases rapidly. Not only the computer anarchists of Hacktic, but banks and insurance companies have raised fundamental objections to encryption legislation in the Netherlands and comparable initiatives in the United States. Even more fundamental is that ICT makes decentralisation possible while simultaneously encouraging centralisation through networks and standardisation. Perhaps even this pair of concepts – decentralisation-centralisation – are becoming meaningless in many respects through the space–time effects of ICT and informatisation. In any case, it is clear that ambiguity is an inherent characteristic of the technology.

Informatisation can be seen as the youngest shoot on the stem of modernisation. ICT is the embodiment of functional rationality: techniques and procedures guarantee the quality of information and communication and make its actual contents contingent. In any case, dominance, as the real goal of every technology, reaches out from information technology to man and his social relationships by substituting technical for intellectual capabilities (Frissen, 1989 and 1991).

However, dominance is not a one-sided relationship. It is not merely a question of one powerful individual dominating the many who are powerless. Domination is a characteristic of a complex set of social relationships within which the one dominates and the rest are dominated. The imprecision of this domination is also connected to informatisation.

On the one hand, informatising organisations will strengthen their control over the social environment. On the other hand, the information-technological equipment available to the actors in that environment also increases. There is no question of one-sided dependency but an increase in mutual dependency in which information systems and communication infrastructures play an increasingly important strategic role. The outcome of negotiations, exchange processes and power conflicts in such circumstances is not only contingent but also ambiguous and its evaluation will depend heavily on one's position in the arena.

Informatisation not only creates ambiguous dominance, but it also generates a far from straightforward process whereby on the one hand complexity is increased and on the other relationships become fixed (Frissen, 1991: 12-13). Information systems when applied to large-scale transactions establish a fixed system of regulation and policy implementation.

At the same time, however, we find historically that the availability of information technology in policy-making sectors has greatly increased the complexity and variety of these systems of regulation and implementation. The intermittent nature of this process may possibly lessen as greater intelligence is built into the system. But even then ambiguity will remain so long as the learning ability of computers lags behind that of humans and organisations.

Analogous to alternating complexity and fixation is that of innovation and rigidity. On the one hand, it is argued that informatisation can contribute to the renewal of organisation structures and processes. Organisations can become flatter, the hierarchy can be replaced by communication networks, functions can be enriched. But against that one can set an element of rigidity. Information technology is introduced into organisations as a 'reinforcement strategy' (Hoff & Stormgaard, 1990) for the dominant actors. A study made by Danziger *et al.* (1982) appears to confirm that informatisation in organisations serves the ends of powerful actors and thereby strengthens existing power relationships. The effect is also observable: because of its strategic importance and the large investments associated with it, information technology becomes an instrument of prevailing objectives and relationships. The context of application and usage can by no means be regarded as a 'level playing field' on which the players enjoy equal powers.

There is a further ambiguity which is an extension of this: that between uniformity and variety. The current state of affairs in the field of technology obviously tends towards uniformity. Standardisation, formalisation and harmonisation of information systems and communication infrastructures are technical requirements. Links between information systems are not unwanted side-effects but fundamental objectives. Furthermore, the large scale of many information systems inevitably leads to a centralisation of authority. Current concerns about security only strengthen this tendency. Uniformity, therefore, is a logical outcome of technological development.

But within this tendency towards uniformity, the technology stimulates a parallel move towards greater variety. If organisations dispose of a number of basic registries and an advanced communications infrastructure, the decentralised or local development of innumerable applications becomes possible.

In this respect, networks are of great importance: organisations and actors within organisations can use the same data for different purposes, and furthermore also have access to each other's systems. This leads to increased variety within and between organisations on the basis of greater uniformity in the communication infrastructure. Here too, both effects are indissolubly connected. In consequence, organisations will simultaneously grow and shrink in scale. Thanks to ICT, the centre can extend its span of control while at the same time permitting far-reaching decentralisation (Snellen & Van de Donk, 1987). Organisations can more easily develop the characteristics of an archipelago because control and governance of inputs and outputs are technologically simplified. Communication between the islands of the archipelago can be stimulated and guaranteed by the communications infrastructure.

I have here sketched only some of the ambiguities: dominance-autonomy, complexity-delimitation, rigidity-renewal, uniformity-variety and centralisation-decentralisation. Ambiguity is inherent in the process of informatisation because ambiguity, contradiction and dialectic are essential aspects of social reality. Social relationships are always the accidental results of a complex set of structures, processes and actors with ever-changing repertoires (Van der Meer & Roodink, 1991). Since information technology is a social artefact and deeply entrenched in our intellectual and political aspirations, this ambiguity also qualifies as a fundamental hallmark of informatisation (see Frissen, 1992b: 189–192).

4 Public Administration and Technology

4.1 INFORMATISED BUREAUCRACY: AN IMPRESSION

It has frequently been observed, but it bears repeating, that information and communication technology has long been widely applied in public administration. Adverse comparisons with private industry and stories of frequent and disastrous computer failures more generally issue from an anti-government neo-conservative ideology than from systematic observation or research. Governments invest huge amounts of money in informatisation at every level. For example, the Dutch government spends 2 billion guilders per annum (Zeef, 1994: 3) while in Britain, 1.7% of public expenditure, or over £2 billion per annum is spent on informatisation (Willcocks, 1994: 13).

Within Dutch public administration, a wide range of different systems and applications are in use. And the Dutch are not alone. Frissen *et al.* (1992b) have published a comparative survey and inventory of the situation in several European countries. The English situation is described in Bellamy & Taylor (1994); the German in Brinckmann & Kuhlmann (1990). In the United States vice-president Gore has stressed the strategic importance of the public sector in creating the 'electronic superhighway'. In Europe, there are numerous research programmes designed to stimulate ICT. So the connection between public administration and technology is obvious. Furthermore, we know from the work of scholars like Foucault that technology, power and politics are interwoven and that every technology is a cultural artefact in the tradition of modernisation and thus of advancing and expanding domination. Technologies can be seen as strategies of discipline, and it is that which makes them so intriguing from the perspective of politics and public administration. All the more striking, therefore, is the relative lack of interest in informatisation within public administration. Hood and Margetts (1993) contend that within the discipline, it has not received any extensive treatment at all. A recent handbook devotes no more than two pages to the question (Palumbo & Maynard-Moody, 1991: 97–98).

Internationally, apart from incidental interest from other perspectives and themes, there are three main research groups at present involved in the subject. In the first place, there is the Irvine group in the U.S. centred on the likes of Danziger, King, Kling and Kraemer. Secondly, there is the Kassel group which has been conducting systematic research for a number of decades and is associated with the names of Brinckmann, Grimmer and Lenk. (See also Brinckmann & Kuhlmann, 1990). And finally there is the Tilburg-Rotterdam research group centred around Snellen, Tops and myself. The findings of this last group (an interim report on the 'state of the art') are published in Zuurmond *et al.* (1994). (See also Edwards, 1993). Contributions from all three groups are to be found in Snellen & Van de Donk (1989).

A first impression of the significance of informatisation for Dutch public administration ought to be given by means of a typology of systems and applications. However, since a universally acceptable typology does not yet exist, I shall start by describing various classifications offered by different writers, and use concrete examples of projects and systems to give some impression of informatised bureaucracy. The examples are merely illustrative of the variety and breadth of the question.

Systems and Applications: Classifications

A well-known typology is that of Brussaard. He distinguishes object systems, sector systems and control systems (Brussaard, 1989: 136).

Object systems contain data about objects, both human and non-human. The best known examples of these systems are the public registers used for judicial matters. Systems containing information about population, real estate, commercial and personal vehicles, social objects, are all examples of object-systems. Brussard also includes public documentary information provision, such as that held in public libraries.

Object systems have their own basis in law and are accessible by public and private organisations. Their most important function is to satisfy external demands for information. Organisationally, there are many different kinds. They exist in decentralised, de-concentrated, centralised and concentrated forms. Whenever organisations use information from different object-systems, this organisational variety can lead to problems of co-ordination.

Sector systems are all those systems which perform functions for a single sector of government. Those functions can vary greatly. They can be basic registries for a sector, transaction systems or systems which support decision-making. It is striking that there is relatively little formal regulation of these systems. Sector systems can supply the information requirements of a range of organisations within a single sector. They function on the basis of similarity and coherence of information within the sector.

Finally there are the *control systems* which operate in a government organisation. They are directed at internal controls and governance and are concerned with such matters as personnel, finance, documentation, accommodation and so on. They can be found in every sector, and across the sectors they display many similar features.

Another typology is that of Snellen and Van de Donk (1989). They distinguish between basic registries, monitoring systems, simulation systems, transaction systems and expert systems.

Basic registries include Brussaard's object systems and also some of his sector systems. They are essentially databases containing data about objects, processes or persons. Normally they contain the minimum information which the various user organisations need to store.

Monitoring systems are systems which follow industrial processes. They can register data about critical events, they can register transactions, they can track human dealings. Their function is primarily for control and management but they can also supply information to support policy processes.

Simulation systems are used to calculate the future effects of alternative policy plans. They can also make use of databases stored elsewhere.

Transaction systems are systems which replace human actions, usually decisions, in an implementation process. Familiar examples are student grants and individual housing subsidies. After importing specific data, the system makes a decision.

Expert systems support decision makers on the basis of expert knowledge and logic structures stored within the system. These systems are being developed in a variety of domains and we may expect to see them applied increasingly in public administration.

Finally there is Snellen's (1994: 285) instructive typology which he bases on various core technologies. He distinguishes:

Database technologies. The introduction of relational databases in particular has released a huge potential for analysing data and combinations of data. Databases are used in public administration for:

- object systems as described earlier
- sector systems, also described earlier
- control systems, which monitor financial, human and physical resources.

Decision-making support technologies. These form an important resource for decision-making through their use of decision-making rules. They can range from relatively simple transaction systems through to advanced knowledge systems, expert systems such as simulation models, and geographical information systems. In public administration, these technologies are used extensively in the 'back office' for administrative and assessment purposes. They are also being used more and more for collecting information through direct interaction with the public. Increasingly this involves connections with linked databases in a wide range of domains.

Network technologies. Within public administration there has been a sharp increase in the use of networks both for specific purposes and for more general information exchange. Standardisation is of great importance in this process (See also Perlee, 1993).

Technologies for personal identification and tracking. Identification numbers, from the very general to the very specific, make it possible to link different databases together and create virtual databases. Smart cards are widely used to provide services and to follow the actions of individuals. Tracking systems will be used more and more in public administration, both in support of such goals as control and service provision, and in tailoring work processes to information flows.

Systems and Projects: Examples

A helpful impression of informatised government can be obtained from the numerous systems and projects investigated by our research project. A few salient examples follow.

GSDs (Local Authority Social Services)
The Local Authority Social Services (GSDs) have been studied at different times by our research project (Scheepers, 1991; Zuurmond, 1994). This prototype of 'street level bureaucracy' shows an extensive and fundamental process of informatisation. A process which began in the 'back office' of the administrative and transaction systems has now moved into the 'front office' of direct service provision, linking with related organisations in the networks of social security and employment policy. Zuurmond (1994: 280–283) describes this process as a 'rationalisation of information provision in cases where the structure of information domains is consciously becoming more functional'. Through informatisation, the organisations become both easier to control and less hierarchical, i.e. more 'horizontal'. Moreover, building on Scheepers' findings (1991), Zuurmond suggests that the integration of information management among the GSDs is also increasing. The physical integration of the mainframe is being succeeded by 'virtual (logical) integration'. Social Security Numbers, in particular, are playing an important role in facilitating the linking of databases. Improved verification, important in combating fraud, comes about in this way. The constraints on implementing social welfare, so often observed in past research, seem to be reduced considerably through informatisation.

Student grants
The project to automate student grants, researched by Van de Donk, Frissen and Snellen (1990), is probably the best-known automation disaster in Dutch public administration. It is also a fine example of the increasingly common practice of synchrony between legislation and system-development: the system which has to apply the law is designed at the same time as the legislation is being drafted. Result: parliamentary debates are effectively faced with a

fait accompli because not only is the legislation on the table, but the system for implementing it has already been designed. The high costs of redesigning the system make changing a law financially problematic. The student grants project shows on the one hand that without informatisation, complex and large-scale welfare arrangements are not possible, while at the same time, in the logical dynamic of system-development, complexity and scale have to be fixed at some moment in time. (See Frissen, 1991: 13) A system which is being built on is already past the point of no return for the legislators. In the case of the legislation on student grants two points arise: on the one hand, complete clarity in the legislation is important so that robust systems can be constructed; on the other, there is an ambiguous tension between simple laws and regulations for the sake of an efficient system, and the potential for complexity and differentiation for the sake of political effectiveness which the availability of ICT encourages.

Inland Revenue
Informatisation plays a crucial role in the Inland Revenue. The large-scale reorganisation of the service in the 1980s was, in a number of respects, facilitated or supported by informatisation. Research by Bekkers and Frissen has shown in this context that informatisation has led to greater efficiency, more effective fraud detection, orientation of work processes and structures towards clients and target-groups, improvements in services and public information (Bekkers & Frissen, 1992: 568–569). More recently, research (Huigen, Thaens & Frissen, 1994) has been done into the development and implementation of an information system which will handle most tax returns automatically. Special parameters within the system make it possible to allocate tax bands, compare tax-return entries and compare categories of tax liability. The system also generates a wide range of control information. Considerations of efficiency, effectiveness and justice can thereby be made increasingly explicit.

GBA (Local Authority Basis Administration)
The Local Authority Basis Administration or 'GBA' is a wide inter-organisational informatisation project. It is a network to which every Dutch local authority and several hundred public and private consumers are connected, over which they exchange information about the population. Data exchange is standardised within a logical framework laid down in law. The work done previously by local registry offices is now conducted entirely via the GBA. The GBA can also function as a key for information exchange between other databases. Local councils and consumers are free to use the GBA network for other purposes. Other computer applications can be linked via the network. The GBA thereby facilitates information exchange between every local authority and a large number of consumers. Our research has shown that the GBA provides a wide range of material as well as, in particu-

lar, non-material advantages in both the inter- and intra-organisational spheres (Bekkers, Straten, Frissen, Tas, Van Duivenboden, Huigen & Luijtjens, 1995).

Local Administration
Local administration and local democracy are important themes in the work of the Tilburg–Rotterdam research programme 'Informatisation in Public Administration' not least because there is such a wide variety of informatisation at this level of the Dutch administration. The following examples, based mainly on Tops, Depla and Korsten (1993) and Depla (1995), are far from exhaustive:

- informatisation to improve the local provision of services, such as civic service-centres, decentralised district services, public information systems and integrated service counters;
- informatisation to improve local policy processes, such as the 'Tilburg Model' initiative, various council information systems, and the welfare index in Enschede;
- informatisation to support policy-making and the setting of social and political agendas, such as community discussions, consultations, surveys and panels.

Digital Cities
Amsterdam's 'Digital City' is an experiment which, building upon the Freenets, brings together many of the aspects discussed above. It is a networked system which combines different forms of information provision and communication and is accessed by modem. The Digital City is also connected to the Internet (see Schalken & Flint, 1995). The Amsterdam model has since been copied by a number of other Dutch cities. Via the digital cities it is also possible to engage in nation-wide political debates. The Ministry of Home Affairs set up a number of discussion groups during its preparation of the BIOS-3 bill (see Zouridis, Frissen & Schalken, 1995). Answers to parliamentary questions about the bill could be found more rapidly via the Digital City than through regular parliamentary channels.

These are just a few examples; many more can be found in Zuurmond *et al.* (1994). They show that informatisation affects public administration at all levels and in all its functions. But before attempting to assess its significance we need to review again the characteristics of informatisation (see §3.3).

70   *Politics, Governance and Technology*

4.2   CHARACTERISTICS OF INFORMATISATION

In the previous chapter I listed some characteristics of informatisation. They are important for public administration. First of all, the characteristics taken from Van de Donk and Tops (1992):

1. There is a sharp growth in the quantity and accessibility of information. Bureaucratic obsession with information is proverbial. It is not for nothing that public administration is often called an information-processing system. ICT offers in that respect near-inexhaustible possibilities. Traditional limits on the availability of knowledge seem to be dissolving, not only because ICT makes more knowledge increasingly accessible but also because its analytical qualities are constantly expanding. The administration's knowledge about citizens, organisations and businesses is steadily growing in quantity and quality. This in turn feeds the administrative love of discipline and intervention. Whether the relationship is bi-directional, giving citizens, organisations and business more information about the administration, is hard to tell. The use of ICT in providing improved information will undoubtedly also be put to the service of tighter controls and combating fraud.
2. As the speed of information provision accelerates, so will immediacy and proximity. Through ICT, public administrators can be informed extremely rapidly about the effects of policy, about public opinion, and about changing patterns in transaction processes. Factors of time and space are less of an obstacle and this has consequences for both organisation and policy-making. Organisationally, physical location and structure become less important because the speed of transmission and feedback is so much faster. For policy-making, functional distinctions between phases of the policy process become more problematical. The speed of information provision not only reduces the distance between policy-development and implementation but also increases the risks if the distance is too great. Van de Donk and Tops (1992: 38) have highlighted the political implications of this: politicians are compelled to react more quickly and more immediately.
3. Receivers of information acquire greater control. Users can decide for themselves when they want to receive information because the range of media and especially the potential for selection is expanding. Public administration will undoubtedly be confronted by the power of the calculating citizen. How effective such strategic behaviour can be, remains to be seen. But it is certain that citizens and organisations will become more demanding about the manner in which they receive information.
4. Information providers can operate more selectively. Information-targeting is already widespread in American politics. As available infor-

mation about the electorate becomes more abundant and more refined, it becomes possible to target specific groups more accurately. A possible outcome of this is a 'balkanisation' of the electorate. It also offers innumerable possibilities in the sphere of policy information or propaganda. Using ICT, custom-made policies can be developed, implemented and communicated with great precision.

5. ICT encourages decentralisation. Perhaps the most important characteristic of ICT is the huge expansion in the capacity of individual systems. This threatens to make obsolete such concepts as centralisation and hierarchy which have been so essential in public administration. Organisations can now become 'virtual' because hierarchical control structures and intermediate positions in the provision of information, are becoming less and less necessary. Decentralised implementation and development of policy are increasingly easy to organise. The necessary information is now available everywhere. I have touched on this earlier.

    However, this characteristic is ambiguous: every argument used for decentralisation can also be used for centralisation. After all, there are no longer any obstacles to proximity and immediacy. A local multi-function counter can be used simultaneously to distribute departmental services, fulfil an ombudsman's function and collect data on policy implementation.

6. ICT is highly interactive. The potential for interaction with citizens and social organisations is increasing exponentially. This means that policy-implementation can be combined with intensive feedback and quality control. For the development and preparation of policy it considerably extends the potential for the co-production of policy. In the domain of political decision- and opinion-forming, it is precisely this characteristic which has 'given rise to wild speculation' (Van de Donk & Tops, 1992: 39). Direct democracy could certainly receive a powerful impetus, though at present the number of experiments in this area is still limited. (See Depla, 1995.)

    The characteristics listed by Bekkers (1993) are also relevant to public administration. As we saw earlier, they focus primarily on the control aspects of informatisation.

Calculation
Informatisation facilitates fast and complex calculations. This has various consequences for public administration:
- In principle, transaction systems can be made increasingly complex by adding extra parameters.
- Increasing quantities of data can be processed in support of decision-making.
- In the preparation of policy, the calculation of alternatives can be increasingly refined.

- In the preparation of policy, new alternatives can be developed, using combinations of data (Bekkers, 1993: 129; see also Frissen, 1994a, 1994b).

The 'calculating state' for which more knowledge means better decisions, is receiving a huge impetus. The symbolic significance of information is being powerfully boosted by technology (Feldman & March, 1981).

Control and discipline
Informatisation is a form of domination and discipline. It is a new strategy in the exercise of power which is being added to the bureaucratic arsenal. Standardisation and formalisation are important bureaucratic characteristics which are also technical requirements for ICT. So these characteristics of ICT spill over into the internal organisation and external policy processes of public administration and affect the citizens and organisations involved with them.

For internal organisation it involves:
- the reflexive power of ICT which leads almost spontaneously to control-information over every domain where it is applied (Zuboff, 1988);
- imposing structure and routine on work processes;
- making information and communication patterns explicit and formal.
-

External policy processes are subject to the following influences:
- Information systems structure, the policy domain and the information and communication relationships contained therein (see Zeef, 1994);
- Information systems make discipline easier by extending monitoring facilities
- Information systems 'define' situations more invasively and transparently: 'the computer is never wrong'.

Transparency
The property of transparency is both a consequence and condition of the characteristics of calculation, control and discipline. In public administration there is increased transparency within the individual organisation and its policy domains. This is a result of the reflexive capacity of ICT, referred to earlier. Next to that there is also an increase in inter-organisational transparency. Networks between organisations make different organisational information systems more accessible.

The same applies to policy domains: as ICT expands so does transparency. At present this primarily affects registration, monitoring, control and verification. Reciprocity is limited: the citizen sees far less of the administration than the administration sees of him. Transparency is not a neutral quality because information and communication are embodiments of bureaupolitical conflict and social power relationships (Frissen, 1991: 13–14).

Communication

The uniqueness of this technology lies in its communicative characteristics and they are becoming increasingly important for public administration whose primary activities – policy development and implementation – are communicative. In other words, technology touches the very heart of public administration. Existing patterns are supported and replaced by technology, creating new patterns.

The communicative properties are creating important opportunities for organisational change: physical proximity and hierarchical control are becoming less important. The network is becoming the new pattern for relationships and creates new forms of inclusion and exclusion (see Snellen & Wyatt, 1993).

More feedback mechanisms can be incorporated into policy processes while possibilities for co-production are enhanced. At the same time, technological competence is becoming more important as a resource for policy networks, which can introduce new asymmetries into the power structure.

How should one evaluate these characteristics and their significance for public administration? Is there a detectable pattern of development? Can one speak of technological determinism by which public administration will be fundamentally transformed, or are there patterns of development which will be determined, at least partially, by the *desiderata* of political and bureaucratic actors? In the following sections I shall attempt to answer these questions against the background of various positions, including my own, which have been taken in the technological debate and were described in the previous chapter (§3.4 and §3.5).

4.3 BUREAUCRATISATION AND TECHNOCRATISATION: THE PERSPECTIVE OF MODERNISATION

'The question arises: how can it be made better, faster and fairer? And then one comes up with automation.' This was how a civil servant, whom I was interviewing, summed up the complex motives which underlay the informatisation of his department (Frissen, 1989: 197). In other words, informatisation contributes to the modernisation of government organisations: efficiency, effectiveness and justice are core concepts in the discourse of modernisation. In this, one can observe a striking similarity between the goals of informatisation and Weber's definition of the ideal-typical characteristics of bureaucracy.

Bureaucratic Culture: an Investigation

Informatisation could very well rival bureaucratisation as one of the most important features of the historical process of modernisation. At least that was the conclusion I drew from an empirical investigation into the significance of informatisation for the culture of government organisations (Frissen, 1989). By bureaucratisation is meant the advance, within a wide range of domains and organisations, of a pattern of organisation characterised chiefly by standardisation, formalisation, specialisation and centralisation.

My investigation, into the General Directorate of a government ministry showed that informatisation, understood as a complex of phenomena surrounding the introduction of information technology into organisations, strengthens the bureaucratic culture of the organisation concerned. The research, which was conducted in the ethnographic tradition, shows that informatisation in the Directorate acquires many of its values and norms from the meanings attached to the term 'information society'. This is fixated on control and information as a condition for success and change. Well-structured and centralised data provision is considered to be of crucial importance. The classical conception of rationality is fused with the technology and technical procedures which lie at the heart of the informatisation process.

In the Directorate this creates a powerful orientation towards financial control supported by information systems. Modelling, precision and quantification are of paramount importance, and informatisation fits seamlessly into this pattern of values and norms. By virtue of these cultural characteristics, informatisation is also able to contribute to a purer form of political decision-making by suppressing irrational and arbitrary elements. More and improved information is expected to improve the policy process. The myth of rational control assisted by optimal information provision is thereby strengthened. Bureaucratic values like standardisation and formalisation appear as technical requirements for system-development, and informatisation becomes the embodiment of cultural characteristics in so far as they contribute to greater legal equality and more rational management.

In the Directorate, the implementation of policy is particularly strengthened by informatisation. In that phase of the policy process, there is a large concentration of information technology. Because of the sensitive nature of information systems and the conviction that informatised processes are more stable and easier to control, there is a tendency to tighter policy-development and political decision-making. Stable policy implementation, supported by extensive information systems, can not tolerate frequent politically-motivated intervention. Technical practicability becomes an ever more important criterion. Thus we observe a shift of power within the policy process itself (Van de Donk, Frissen & Snellen, 1990).

Wahlverwandtschaft

The pattern of values and norms embodied in informatisation is strengthened quite significantly by experts in the field of information technology. They are protagonists of a technology which has few critics and whose function is to bring about and refine rational government organisation. Increasing dependence on this expertise suggests that its norms and values will also have a strong cultural influence on government organisations.

Interpreting research findings leads to the conclusion that there exists an ideal-typical *Wahlverwandtschaft* between bureaucratic culture and the patterns of meaning embodied in information technology and informatisation.

Standardisation and formalisation are, as already stated, technological requirements for effective system development in public administration, as well as normative demands and value categories in respect of organisational processes. They regard stability as a desirable aspiration and instability as evidence of slovenliness and loss of control. The greed of the technology results in a strategy of discipline: ongoing standardisation and formalisation, and the combating of ambiguity and uncertainty. Through informatisation, functional rationality (a pattern of meanings in which the concept of 'goal-means-means' is central, and methods and procedures play an important role) is on the advance; substantive rationality (in which norms and values are accorded the central role and, in the context of public administration, is represented by politics and ideology) is losing ground (Frissen, 1991: 8–9). So one can speak of a technological culture, in which it is not just the context in which information systems are used and developed which counts, but more especially the cultural characteristics of the technology itself.

Technocratisation

In public administration the need for progress, associated with informatisation, is taken for granted. The metaphor of public administration as an information processing system symbolises this tacit consensus (Zinke, 1990). Information, generated by machines and controlled by technological procedures, is the crucial element of what Beniger (1986) calls the 'control revolution'. Technology extends out into control and discipline. The unique, indeed the radical nature of informatisation is that control is no longer directed at the physical environment but at ourselves in our social relationships. The pretension of informatisation is self-control with the help of an intellectual technology (Bell, 1979).

The intellectual character of the technology also explains why bureaucratisation, as the main consequence, needs closer specification. Bureaucratisation here has a technocratic meaning. In other words, the process of bureaucratisation brought about by informatisation in government organisations is an advanced process in which a more (techno)-scientific rationale becomes the interpretative framework, and expertise is embodied increasingly in configurations of systems and professionals.

> 'It is (...) the unfolding of this system of intellectual and organizational rationality that brought the technocratic project to the fore. (...) The technocrat, in short, is assigned the central task of fitting the bureaucratic organization to the technological mission of modern society.' (Fischer, 1990: 63-64)

What Fischer, however, does not sufficiently recognise is that the technology is no longer external in respect of the technocrats and that it is not just a matter of applying instruments to the achievement of goals. The really revolutionary aspect of ICT is that the technocracy and the technocrats are being replaced by technology. Technocratisation is no longer the consequence of subjects' intentions, but is an institutionalised framework of meaning within the dominant technology.

The Infocracy

The above conclusions seem not only to be confirmed in a recent study by Zuurmond but, in a sense, are taken even further. In the light of developments in the GSDs, he argues methodically and logically that informatisation is an advanced form of modernisation. Thus it is not a matter of '*Wahlverwandschaft*' between bureaucratisation and informatisation, but rather of informatisation being used to elevate bureaucratisation to a higher plane. Where bureaucratisation attempts to achieve greater control through the specific structuring of organisations, the structuring leads inevitably to a failure of control (Zuurmond, 1994: 274). Bureaucratisation appears to come up against in-built boundaries of a physical, cognitive, political and social nature. Informatisation, according to Zuurmond, is the answer to this loss of control as well as being an attractive alternative to the popular demand for less government. Informatisation is a strategic reaction to the crisis of bureaucratic control; it can lead to more effective control while appearing to adopt a strategy of de-bureaucratisation.

Informatisation creates an information-architecture which takes over bureaucratic control and co-ordination and makes them invisible. Formalisation and standardisation appear to disappear from the organisation, but are *de facto* taken over by the systems. Though ICT is introduced because of the size, indivisibility, complexity, interwovenness and dynamism of the policy sector, its application makes the organisation appear to be more horizontal, less hierarchical and more flexible. Zuurmond's conclusion is that Weber's ideal-type needs to be updated. The new ideal-type of infocracy has the following elements (Zuurmond, 1994: 284–287):

- As regards its framework of meaning (Zuurmond speaks of its 'norms'), infocracy is a realisation of the following goals and values: continuity, effectiveness, obedience, calculability, efficiency, integration, virtuality, indivisibility, controlled complexity and rapidity.
- The instruments of infocracy are partially the same as those of bureaucracy, though modified and partly new. Centralisation particularly affects the authority to use and create data. Hierarchy is considerably refined: as well as sequentiality there is also parallism and reciprocity. Specialisation is reduced in general, but is heightened in the area of information architecture. Standardisation of information provision is a pre-condition for doing tailor-made work elsewhere. In absolute terms, there is less formalisation, but relatively it increases especially through data-coupling. Informatisation in terms of its extent and control is total.
- The effects are that: 'The (...) digital domination of infocracy creates a virtual fortress which is even more difficult to penetrate than the wall of rational, legal bureaucracy. It serves an increasingly individualised consumer society which is characterised by great prosperity, indivisibility, high complexity, dynamism and interwovenness.' (Zuurmond, 1994: 287)

The Balance: Informatisation and Modernisation

With this infocracy, which is of course ideal-typical and not empirical, a pattern for continued modernisation is sketched, in which advanced bureaucracy – without the classic vicious circles of bureaucracy (Crozier, 1963; Vroom, 1980) and without its inherent contradictions – engages closely with the technocratic capacities of the technology.

Informatisation, as many observers have argued, is a process which can without any doubt be seen from a perspective of domination and control. Like all technologies, information and communication technology is primarily directed to increasing control over physical and (now) social environments. Informatisation is therefore an authentic part of the modernisation process, a process which can be characterised as the ongoing development and unfolding of technical and scientific rationality.

As far as public administration is concerned, the advance and development of technical and scientific rationality is a process of bureaucratisation. On the one hand, there is an expansion of legal and rational control; on the other, an improvement of the instruments of that control. The autonomous process of bureaucratisation means that the objectives and instruments of that domination flow into each other; functional rationality is simultaneously target and instrument. The political aspects of that domination are thereby put under pressure (Frissen, 1991; Kuypers, Foqué & Frissen, 1993).

Because informatisation is not just a collection of neutral instruments but also a complex of technologies and practices linked to knowledge, it is perhaps primarily a process of technocratisation. Not so much in the sense of technocratic dominance, though that is true, but particularly because technocratic expertise, embodied as it is in information and communication technologies, is becoming increasingly important in public administration. Informatisation is a process in which machines collect, store, process, distribute and construct knowledge; a process in which machines 'think'. The Faustian ambition to create life out of dead matter seems to have long been achieved in a number of aspects. Information technology is an intellectual technology which drives the individual subject from the centre of power, or at least makes a pact with him. The '*Entzauberung der Welt*' is now affecting human beings.

In spite of all voluntarist and constructivist nuances, the hard deterministic core of technological development is thereby revealed. Again it raises important implications for politics, because politics is one of the patterns of meaning of an anthropocentric world picture. As technocratic expertise grows in importance and the human decision-maker disappears from the centre of power, what is left of the primacy of politics or the sovereignty of the public domain?

4.4 INFORMATISATION, CHANGE, EXPERIMENT AND INNOVATION

Zuurmond's research shows that informatisation does not necessarily result in classical bureaucratisation. Less hierarchy, less rigidity, more flexibility, more task integration are but some of the non-bureaucratic consequences of informatisation. But that in no way diminishes the perspective of domination. It appears rather to lead to a more advanced form of bureaucratisation, without its immanent contradictions and with the full realisation of modernist ambitions. Without going into great detail, Zuurmond's argument seems to support the conclusion that the relationship between informatisation and many forms of administrative change, experiment and innovation is problematical. Problematical in the sense that informatisation has to be viewed from a perspective of control. Administrative renewal undertaken with the help of ICT, therefore, is a refinement of bureaucracy. 'Thus instead of Weber's iron cage with the tangible bars of a hard, rigid structure, we have a *virtual fortress*' (Zuurmond, 1994: 287)

Perhaps his conclusions should be taken to apply primarily to the empirical domain of his research, social security, but there seems to me to be room for wider application. In a number of respects, renewal is improvement, in others it is transformation. Often we see both together in conformity with the ambiguous character of informatisation in public administration.

Controls and Tailored Services

Large-scale investment in information technology by the inland revenue, social security, housing benefit, student grants; a striking intensification of verification, testing, control by means of cross-linking databases; the introduction of the GBA; the use of social security numbers; the plans to introduce smart cards into health care; driving on account etc. All these activities and developments can be seen as using ICT to increase control over citizens, organisations and businesses.

It is remarkable how quickly concern about privacy has been overtaken by concern about fraud. ICT is being used more and more to check on policies and regulation, but very few political, legal or social objections are being raised, or at least being heard. Social security fraud is no longer the exclusive theme of (neo-)conservative circles. There is now a broad social consensus on its prevalence and the need to combat it. On the one hand, the change in climate has stimulated the application of ICT as a control mechanism; on the other, this use of ICT has heightened the awareness of fraud. In this respect, there is no question of renewal but rather a clear continuity in the disciplinary and control ambitions of the modern (welfare) state. Over and against the 'calculating citizen' we see an increasingly well-equipped and politically less scrupulous monitoring state (Frissen, 1994a; Van Gunsteren, 1994).

Set against that, or better, closely connected in the sense of being the other side of the coin, are the service ambitions of the modern state. One should naturally be sceptical about the current market concept of the citizen as a client of government. Monopolists and clients do not sit comfortably together. Nevertheless, one implication of the market concept is that bureaucratic logic should no longer be at the centre and that a service ethos should be allowed entry into public administration.

The citizen, then, is independent and must be treated like an 'adult'. ICT offers a number of possibilities:

- The quality and quantity of information provision can increase
- Information can be concentrated at single points via integrated service counters
- Service provision can be integrated
- Tailor-made services can be offered
- The lack of a true market mechanism can be compensated for by fast feedback mechanisms.

The image of government (Ringeling, 1993) is thereby enhanced by ICT; the extensive research of Tops, Denters, Depla, Van Deth, Leijenaar and

Niemöller (1991) shows that the public also perceive it in that way. This does not indicate any revolutionary change but a gradual improvement in the quality of service provision. Efficiency, speed, flexibility and tailoring to specific needs would be virtually impossible without ICT.

Anticipation and Ambitions

Informatisation also opens up various possibilities of anticipation by the administration. I use anticipation in the widest sense: all forms of policy preparation and development, as well as strategic decisions. In other words, every attempt to deal intelligently with the future.

In the first place, there has been considerable growth in the information collected about social activities. That occurs in the databases and information systems of the administration. Access to external databases is also becoming easier. Each increase in the amount of information can lead to new policy intentions and initiatives, and experience has taught us that this does often happen.

In the second place, it is not just a matter of a quantitative growth of information, but also a considerable expansion of processing potential. More databases can be combined and compared with each other. More analyses can be carried out and the results made almost immediately available. Such analyses can lead to a diagnosis of trends, of discontinuities, of profiles, of accumulation; in short, of all kinds of patterns and phenomena which themselves become the input for further policy plans and initiatives. Technological development is making it possible for systems to offer these facilities automatically. Apart from all this, information about citizens is increasing because of the growing quantity of electronic traces which they leave behind them.

In the third place, public administration is making more use of ICT-based simulations and exercises (see Geurts, 1993). ICT is becoming an instrument in the processes of policy preparation and development. Decision-making situations are being supported by different systems and applications. It is evident that this too has policy implications when we consider the influence which official calculations have on consensus-building in periods of cabinet-formation.

ICT's stimulus to anticipation is not confined to any specific phase of the policy cycle. It can be in the preparation as well as in the implementation. Furthermore, one of the interesting effects of informatisation is that both phases of the policy cycle can be more closely dovetailed because of accelerated throughput and feedback of information.

In policy preparation, ICT's effect on anticipation can be seen in the following:

- agenda forming through processing and coupling of databases (Meyer, 1994);
- policy development through analysis, and through simulation and exercise techniques (Geurts, 1993);
- policy choices, including allocation policy, with the help of decision-supporting systems.

In policy implementation, ICT's contribution is in generating information and evaluating policy through the use of systems and applications.

By expanding the possibilities of anticipation, policy goals are stimulated. Lack of information is in principle no longer a limitation, or, at least, the limits have been greatly extended by the sharp increase in the quantity and quality of available information. This may inspire better discipline and control as well as improved service provision. It is even probable that these will go hand in hand.

Independence and networking

Business literature is full of claims to the effect that ICT makes greater branch autonomy possible. Profit-centres, business units, 'results units' are given more independent responsibility. Keeping a check on them is facilitated by electronic networks and on-line monitoring. But curiously, this reasoning is rarely encountered in respect of independent units in government (see De Vries & Korsten, 1992; Frissen *et al.*, 1992a). Although independent agencies are a prominent feature in theory and practice, indeed it is one of the most important modernising strategies in Dutch public administration, informatisation is not regarded as a strategic factor in the process (Frissen *et al.*, 1992a: 80-81).

Nevertheless, information provision is crucial. The relationship between an independent unit and the mother organisation (the department for example) assumes, and for the most part usually is, an information relationship. The possibility of giving political direction to autonomous units is intimately tied up with the information relationships of governance, answerability and feedback. The specific regulation of competence – the formalisation of power relations – directly involves a pattern of information provision and communication.

The availability of ICT extends the possibility of allowing greater independence to government organisations without diminishing information exchange and communication. And for reasons of flexibility, adaptability and greater participation, there are strong arguments for giving government organisations greater independence and decision making powers.

In a general sense, therefore, every form of decentralisation is also a form of independence. And thanks to informatisation, all those forms are easier to organise.

At the same time, precisely because of its strategic position in the process of independence, informatisation constitutes a threat. Its extensive monitoring facilities can easily be used to limit the very autonomy which independence was designed to create, and without using classical bureaucratic instruments. This lies at the heart of Zuurmond's argument as we saw earlier (Zuurmond, 1994; see also Van de Donk, Frissen & Snellen, 1990).

The relationship between administrative modernisation (independence) and informatisation – which is both facilitating and threatening – remains within a hierarchical or vertical perspective. But the process of granting independence to government organisations, territorially and functionally, can also lead to a powerful expansion of horizontal relationships. Such relationships – with other independent agencies, with actors in the immediate vicinity, with interested parties – are made possible through network technology. Direct information exchange and the shared use of databases create a parallelism between policy networks, social networks and electronic networks. The outcome is a horizontal pattern of relationships.

Of course, this by no means excludes the possibility of central control. Who is to control the communications network? Who will define situations by demanding data definitions? Who will regulate access to the network? Technologies, too, are fixed power relationships. Perhaps some scepticism towards the often exaggerated claims by management gurus is appropriate.

Democratisation

ICT has always generated highly-charged expectations: not the least of which is that it would contribute significantly to direct democracy. Even some trade-marks (Apple) are a direct product of sixties idealism which believed that for the first time technology would not primarily serve the power elite, but the individual – though admittedly not the under-privileged nor sections of the anti-culture (Roszak, 1986: 161 ff.). Since then these voices have not been silenced; on the contrary they have gained new impetus from the developments described in §2.2.

Van de Donk and Tops provide a helpful overview in which the following aspects are distinguished:

- ICT strengthens the potential for direct democracy: limitations are removed; opinion polls are always more reliable than elections; participation, especially at the local level, can increase.
- ICT enlarges the responsiveness of representative democracy because politicians can follow the preferences of the electorate more easily;
- ICT fosters active citizenship, because interactivity is one of its most important qualities (Van de Donk & Tops, 1992: 35–36).

The various actors in a parliamentary democracy are affected by ICT. The importance of intermediate organisations such as political parties may suffer: communicative relationships, after all, become much more direct. The citizen becomes more transparent and more fragmented. The electorate 'balkanises' as a result, and politicians become 'dealers in majorities'. Parliaments may gain more control, but ideological disagreements become more explicit and compromises more difficult (Van de Donk & Tops, 1992: 36–37).

ICT has been introduced in numerous experiments and initiatives for administrative renewal (for a survey, see Depla, 1995). The potential for citizen participation has been greatly increased. Furthermore, in such projects as the Digital City we can see the emergence of political activity from the dynamic of network communication rather than from the logic of the political system: spontaneous discussion, non-territorially determined participation, debates without intermediaries. Whether this is participation in fancy-dress or a genuine change, only time and further research will tell.

Nevertheless, it can be accepted that different characteristics of the technology will have a fundamental effect on democracy. In particular, the rapidly expanding potential for interactivity is likely to become a part of every phase of political opinion and decision forming. The predominantly one-way traffic of the top-down approach will become much less important. Also the disappearance of space–time limitations is of the utmost importance. The 'territory' as focal point for political decision making in democracy is being side-lined now that the White House and the American president can be reached by e-mail.

4.5 TRANSFORMATIONS

The connection between informatisation and administrative change can be seen as advanced modernisation, as continuing differentiation, and as improvement in the politico-bureaucratic system. ICT then is a set of tools at the disposal of bureaucracy and its masters. But it is also more: ICT can be seen as a strategic factor which leads to an increasingly 'sophisticated' bureaucracy, which is less and less the instrument of political rulers, and much more of an autonomous institution. Its inherent dynamic is strengthened and perfected by advancing informatisation from which an increasingly fine-meshed control- and steering-apparatus emerges. Political and policy preferences take on technology's codes of meaning – functional rationality – rather than the guidelines created by 'traditional' decision-making processes based on the primacy of politics.

But there is also a less linear development. The connection between administrative change, experiment and innovation on the one hand and technological development on the other, signals other transformations of which I see four: horizontalisation, autonomisation, de-territorialisation and virtualisation.

Horizontalisation

As we saw earlier, technological developments mean the following for organisations:

- Information can be shared more easily and made more widely available through networks and databases.
- E-mail facilitates communication within and beyond units and hierarchies.
- External sources of information can be consulted more easily.
- Electronic data interchange (EDI) encourages communication between parts of organisations (Peters, 1992:122)

In a number of respects, this has the effect of horizontalising the pattern of relationships within and between organisations. This tendency is strengthened when combined with other administrative developments.

Intra-organisational horizontalisation in public administration can be seen in a number of developments. In the first place, organisations become flatter because the number of hierarchical levels can be reduced: informatisation makes intermediate functions in the provision of information less necessary. This is further encouraged by attempts to reduce the size of government organisations, to work with more flexible units (project-orientation) and to implement network structures within the organisations.

In the second place, horizontalisation comes about because the opportunities for horizontal task-integration are expanding and there is less need for vertical task specialisation (the separation of implementation from supervision). An important impetus in this respect is the ideology of client-orientation. Multifunctional service provision to the citizen – via a service counter, for instance – requires the integration of tasks at the level of the front office as well as an expansion of decentralised decision-making powers. Of course, a centrally defined framework and extensive monitoring facilities still remain in place (see also Zuurmond, 1994).

Thirdly, horizontalisation is on the increase through the growth of informal communication via networks and e-mail. (See Feldman & Sarbough-Thompson, 1993). This will undoubtedly have a knock-on effect in public administration, though as yet there is little or no research data available.

Even more important than the intra-organisational effects is the inter-organisational impact of horizontalisation. Firstly, we see a sharp increase in the number of networks in public administration. The Local Authority Basis Administration network (see above) is a good example. Horizontal information provision and the shared use of data communication facilities between different government organisations at the same or different levels considerably weakens the three-layer pyramid of public administration.

Secondly, in administration we are seeing a development towards partnerships and contractual relationships not only in the public sphere but also between public and private actors. Such partnerships already have a horizontal character, which is strengthened by the common use of networks, data stores and information systems. Task-related activities can be organised more flexibly and in less rigid hierarchies. Existing administrative units take on more of a recording function, while many activities of a temporary nature are brought under temporary structures. The provision of information is being made easier by informatisation.

Finally, the most interesting feature of all this is the combination of intra- and inter-organisational horizontalisation. The boundaries become greyer: 'inside' and 'outside' are alien concepts in a cyberspace of electronic connections. Physical organisation becomes less important than the constantly changing patterns of relationships created to satisfy the needs of specific tasks. ICT is, therefore, not just a convenient instrument, but is itself a set of organisational patterns which transcend existing boundaries.

Autonomisation

Decentralisation, dispersion and independence are administrative trends which have important parallels in technology. As individual systems and configurations become ever more powerful, so individual organisations, their subdivisions and even individuals themselves are becoming better equipped. There are no longer any technical obstacles to decentralised autonomy. Necessary information can, literally, be organised at will. Technological capacity, and especially the integration of different types of technology, makes the centralisation of the mainframe era superfluous. In combination with informatisation, bureaucratisation no longer means vertical integration with hierarchical controls. Informatisation really can lead to 'empowerment': the enlargement of decentralised autonomy. In part, this makes the current debate about the expanding scale of public administration obsolete: just think of education, health and regionalisation.

Technology has made the small-scale possible. One can go further: many of the arguments for large-scale organisation, especially in relation to information provision and communication, are no longer valid. Autonomy is technologically achievable. And in both the practice and the rhetoric of decentralisation, dispersion and independence, we see the emergence of autonomy. ICT makes autonomisation possible in three ways:

- With ICT, decentralised units are better able to collect and process relevant information.

- ICT can create connections between autonomous units without requiring a single physical organisational structure.
- ICT creates autonomy, while at the same time leaving functions such as checking and supervising intact for other organisations (Zeef, 1994; Zuurmond, 1994).

Deterritorialisation

Technological development is creating a situation where the whole world is available on-line on a single lap-top. As time and space cease to be limitations, they become almost irrelevant. The world is not only close in immediacy, but in all its depth and breadth. A little time spent surfing the Internet will illustrate this in concrete fashion. This brings about changes in organisational patterns which were originally set up to overcome the obstacles created by space and time. Organisations organise processes over a specific period for a specific territory. Decision making and control patterns are designed sequentially in linear time-scale. And this is equally true in public administration. Changes in the processes usually aim at improving their space–time performance. Greater independence for organisational subdivisions can be seen as an attempt to achieve shorter response times in terms of feedback between policy implementation and its effects. Regionalisation is an attempt to adapt the administrative structure to pre-empt problems. Technological developments are, however, leading to de-territorialisation because neither time nor space impose limitations. The world is a village – a village is the world. Furthermore, the available information and communication technologies make it possible to organise information-provision and communications virtually independently of existing administrative patterns. Connections and links can be introduced at every level of decision-making and this creates tensions with the existing politico-administrative structure which is organised territorially.

The notion of territory has always had great organisational, administrative and political significance. ICT reduces this because:

- the location and object of data collection can be separated: Dutch citizens could be monitored from New York or New Delhi;
- distributive data need not be confined to a single organisation or geographical domain, as is shown by the Air Miles programme;
- networks enable the creation of temporary communities: President Clinton can be approached simultaneously from Tilburg and Moscow.

Boundaries and processes of exclusion and inclusion do not disappear, but they are far less territorially determined. All this brings into question the hierarchical layering of public administration based on location and territory.

Developments in ICT are now being added to ongoing processes of regionalisation, internationalisation and functionalisation.

Virtualisation

Virtual reality is a reality which is imitated, as in a flight simulator or amusement arcade. This reality can be 'true' or realistic, or it can be artificial and surrealistic or hallucinatory.

In public administration, VR can be applied to a wide range of activities and processes. Obvious examples are the environment and town-planning. One could imagine all kinds of ways to involve citizens, groups and organisations in the design and development of the environment. In a virtual reality machine, citizens could be placed within an actual or imaginary urban setting. The spatial and visual consequences of their preferences would then be made immediately evident. In this way, interaction between administration and citizen would gain in intensity.

Even without the three dimensional facilities of VR, one can imagine the creation of new realities. By linking data files, integrating media, and combining information systems, there are possibilities for simulations, policy exercises, scenarios etc. Also, in a more accidental way, the manipulation of information can lead to the construction of a new reality: policy options, problem definitions, the accumulation of backlogs are all constructions which can be created through ICT.

However, virtualisation can also be applied metaphorically, in particular in organisations involved in changing experiences, or social and physical relationships. Virtualisation of social, organisational and administrative relationships is one of the most important consequences of ICT development. To review briefly the main aspects of virtualisation (see also §3.7):

– simulation of existing realities by integrating technologies;
– creation of new realities by such integration;
– the perception of simulated and created realities as realistic through a type of man–machine symbiosis;
– reducing distances in space and time, because on-line connections over large distances can suggest proximity and immediacy;
– reducing territorial limitations on transactions and communication;
– the design of communication and information relationships at desired levels of scale, participation and information provision (Frissen, 1994b: 443–444).

The boundaries of government organisations, which always designate specific administrative or bureaucratic identities, are made more relative by virtualisation. The connections between government organisations do not just become more important, but they are perceived increasingly as alternative organisational patterns in their own right. If organisations X,Y and Z work together using a network, and they obtain information about transactions from executive organisation A, and link that with information from government organisation B, so that private organisation C can produce statistical analyses which are permanently on line, then in fact the actual organisation becomes the network of informational links and connections which is physically untraceable.

Information and communication relationships, rooted in an infrastructure designed for information-exchange and communication have become the new organisational and administrative format. Moreover, they are, or can be, more fluid and flexible because they are in a sense indifferent to existing physical organisation. So it is technological nonsense for the Secretary of State for the Arts to 'move' physically to the Ministry of Education, Arts and Sciences, just as it is to create new 'city provinces'.

Furthermore, virtualisation will lead to the meanings and situations generated by ICT becoming an increasingly authentic input into the actual policy process. Meyer (1994) argues that this is already the case with the relatively simple databases currently being used. As the quality of the constructed reality improves – i.e. becomes more 'realistic' – there will be a corresponding increase of influence on the policy process. Citizen profiles can be generated from the electronic traces which they leave behind them in innumerable data files. New 'target groups' and more intensive 'information-targeting' then becomes possible. Even a totally individualised approach is not inconceivable.

All this will be accelerated as the feedback time of implementation- and evaluation-information to the policy makers is reduced. Electronically perceived and evaluated reality can then be increasingly finely-tuned and responsive.

It is this virtualisation which, together with other effects discussed earlier, are illustrated in Case 3.

4.6 CASE 3

In the case of independent agencies, all the previously sketched tendencies are shown in condensed form. Independent agencies fit in to the Dutch tradition of (neo-) corporative amendments to bureaucracy and are therefore part of the continuous process of change in Dutch public administration. But independence in public administration involves both modernisation and transformation, and in both cases informatisation plays an important empirical role.

Informatisation as a Strategic Factor

In Dutch debates about independent units informatisation is the forgotten factor – both in theory and practice. Exceptions are Kuiper (1993) and Frissen *et al.* (1992a). There is no mention of the fact that, thanks to ICT, subdivisions of organisations can now be physically situated far from any central direction. Even the deliberations of the Scheltema and Wiegel Commissions (Commissie-Scheltema, 1993; Commissie-Wiegel, 1993) in which independence played such an important part, largely ignored the connection.

The opposite holds for the business world where technology is seen not merely as facilitator but often as a strategic factor in setting up self-sufficient decentralised divisions (see e.g. Peters, 1992; 105 ff.) Nevertheless, in the actual practice of creating independent administrative agencies, informatisation is crucial for a number of reasons.

Informatisation and Independence: Autonomisation

Most government organisations which have recently become independent, or for which independence is being considered, are responsible for the execution of policy. There is a popular assumption that policy development and implementation ought to be kept separate so that, on the one hand, steering can be retained along the main lines established by core departments, and on the other, implementation can be carried out in a relatively stable context under professional management. Executive departments, however, are usually heavily informatised with large transaction systems at their disposal. Independence therefore takes the form of autonomising these departments:

- Technological capacity makes it possible.
- Autonomy guarantees some degree of stability of implementation.
- With ICT, the monitoring of execution and implementation is guaranteed.
- With ICT, feedback of information on implementation is guaranteed.

Informatisation and Independence: Horizontalisation

Independence for executive departments enables them to develop their horizontal relationships. That can be as much the reason for their independence (drawing in social actors as happens, for instance, in employment policy) as the effect (when coalitions are formed in the interests of policy-development). Horizontalisation occurs through:

- participation of social actors in the steering of independent units;
- co-operation between independent units;
- co-operation of the independent unit with public and private actors.

Networks and shared databases play an important role in this context; in particular, the coupling of databases as, for example, student grants or social security with taxation records (see Mieras, 1995).

Informatisation and Independence: Deterritorialisation

Independent agencies are an attempt to bring executive departments 'closer' to the field of operations. That is to say, shorter response times are needed between policy implementation and its effects, between the implementing organisation and the recipients. Independence is therefore a strategy whereby the need for small-scale flexibility and client-orientation goes hand in hand with the control and supervision of implementation.

At the same time, it might be argued that informatisation dilutes the concept of 'scale'. Distributed information makes small-scale provision possible within a large-scale framework. It also reduces the importance of time: feedback of policy outcomes can now be on-line. Territory therefore becomes less significant. Thus the impact of informatisation for independent agencies is that horizontalisation and autonomy become more important, but spatial considerations (with dispersed units, for example) become less so.

Informatisation and Devolution: Virtualisation

The developments which we have been tracing lead to an enhanced virtualisation of public administration along lines very similar to those described by observers of industry (Ettighoffer, 1993; Malone, Yates & Benjamin, 1987; Schmid, 1993). Independence leads to the following configuration of organisational, administrative and informational arrangements:

- Discrete tasks in the policy process are handed over to separate organisations which operate relatively autonomously.
- These organisations work together, more or less permanently, in policy networks which are supported increasingly by electronic networks.
- The tasks are carried out via distributed information resources (virtual databases).
- Core departments will mainly have a directing function or meta-steering task. (See Frissen *et al.*, 1992: 86ff.; Van Twist & In 't Veld, 1994 & 1995)

In this way a 'virtual organisation' of autonomous organisations comes into being, contractually and administratively connected in a network of policy-orientated and informational relationships, and built around distributed information which makes possible specific tasks of a more or less temporary nature.

4.7 THE PRIMACY OF POLITICS

In the discussion of independence in particular, and administrative and technological development in general, little attention has been paid to the political implications. Although they will be the central theme of our next chapter, they merit some attention now since it can be argued that the special combination of administrative renewal and technological development has fundamental consequences for politics and the primacy attributed to it.

The political governance of independent units must be conducted through information and communication relationships that are generally informatised. Furthermore, the scope of ministerial responsibility is seen and often defined in law in terms of those relationships. This means that political relationships are embodied in networks and information systems.

However, if the executive arm is technologically better equipped than its parent department this can easily lead to an imbalance of power. That applies, *a fortiori*, when an independent unit in its horizontal relations with other independent units is better able to generate policy-relevant information. It challenges the assumption that granting independence to executive branches will make it possible to differentiate conceptually and organisationally between policy development and policy implementation. This is not only theoretically problematic, because it assumes a linear and hierarchical connection between determining objectives and resource allocation, but it also becomes irrelevant since technology makes it possible to integrate policy implementation and development, as it were, 'on line'.

Wherever informatisation is applied to policy-making and development, we can see that it is possible to create a direct link between the development of policy and its implementation. The organisational and administrative logic of a division between policy-making and implementation, and the associated cognitive view of policy-making as a reflexive hierarchy (Hoppe & Edwards, 1985) is undermined by informatisation. It is technologically possible to link policy-making and policy-practice closely together and with the help of ICT to keep development and implementation at one level.

Primacy is then no longer an exclusively political matter but becomes social; a tendency which we have seen in other developments. A number of aspects are:

- On-going rationalisation of bureaucracy and the contribution of informatisation lead in a number of respects to a lessening of political ideology as a normative code. Technocratic discourse is given a substantial technological basis.
- Localisation, functionalising and autonomy lead to a further differentiation of the politico-administrative system which can weaken political primacy at the centre with its notions of integration and coherence.

- A stronger orientation towards self-steering by relatively autonomous social sub-systems and an increase of contractual, as opposed to regulated, connections causes a comparable weakening of political primacy.
- In contrast to the strongly horizontal and networked nature of social relationships the politico-administrative system is hierarchical and pyramidal – it is thus inadequate.
- Social complexity leads to circular policy processes in which implementation and strategy formulation become intertwined. This too affects the primacy of politics.
- Informatisation categorises comparable effects as processes of administrative renewal: horizontalisation, autonomisation, de-territorialisation and virtualisation. Politics in the sense of ideology, or in its institutional form of parliamentary democracy, can thereby become marginalised.

Through these developments, the position and functions of the political system become problematic. The political system, whether based on territory, decision-making or normative claims, is just not varied enough to correspond to social reality. Its three-layered structure does not fit the network character of social decision-making. The exclusivity of representative decision-making can not adjust to social primacy and alternative forms of participation. Its normative claims clash with the reality of a de-centred world in which values are relative to each other and the 'grand narrative' can no longer be told. I shall develop this further in the following chapter.

4.8   DIGITAL AMBIGUITIES 2

As always, social development in the real world is neither linear nor straightforward. Ambiguity is its most important and intriguing characteristic. And that is also true of informatisation and its impact on public administration and the primacy of politics – indeed it is true of informatisation in general. Are we here dealing with an instrument, an element of coercion or a construction?

Every effect, every relationship is affected by this fundamental ambiguity. A few of the more striking examples are, for instance:

Links between information systems are intended primarily to combat the unauthorised use or misuse of government resources. They serve the struggle against fraud and are now widespread in Dutch public administration (Van Duivenboden, 1994: 398 ff.). The same links are also needed to improve the provision of services to citizens. Welfare and control are inseparable in the welfare state (Frissen, 1994b). Through ICT, government organisations can be designed from the perspective of the 'clients', the citizens and their preferences. Multifunctional service counters, government service centres, require integrated policy implementation. This is made possible by ICT.

Informatisation is still predominantly a feature of the executive departments. It is there that one finds the large-scale transaction systems for payments, taxes and services. Informatisation makes possible the continual refinement of these transactions.

At the same time, ICT generates information about policy-implementation which can be used to develop new policies. Informatised executive departments provide better and faster information about the effects of policy, about the trends and changes affecting the policy domain, and about the groups affected by policy.

In the sphere of policy development too, ICT is on the increase. Software for simulations, scenarios and decision-support is increasingly available making it possible to steer policy implementation more effectively.

Departmental independence and informatisation have made the relationship between policy development and implementation an exciting and, as yet, uncertain area.

Executive departments can be turned into the payment and service offices of government, while core departments, in providing support even down to the level of programming, can control and influence the implementation of policy (see Zeef & Zuurmond, 1994; Zuurmond, 1994). Policy development and steering thus remain centralised, but supported by implementation information.

A development towards more powerful implementing bodies which, because of their technological headstart, also develop their own policies, is equally possible. This becomes even more probable when they develop horizontal relationship patterns with the help of network technology and shared information systems and databases. Discussions and policy integration can take place away from the centre, without any political or administrative directions. The flexibility of temporary 'work-groups' for specific social and administrative issues provides the potential for this development especially now that most of the peripheral technological obstacles to flexibility have been removed (see Frissen, 1994b).

There is a comparable ambiguity in the relationship between government and society. On the one hand, informatisation provides a powerful impetus to domination and control. On the other, it encourages the autonomy of citizens, organisations and businesses.

In any case, we can conclude from this that although the tendency to control is an important cultural characteristic of information technology, such control involves and serves the interests of different actors by being directed at different objects: bureaucracy and parliament; state and citizen; government and action groups; men and women; democratic and fundamentalist parties. It is not immediately apparent who enjoys power in respect of informatisation and in Foucault's conception of power it is even illogical to think in such dichotomies. Much more important is the actual power relationship in which both perpetrator and victim find themselves at opposite ends. From that perspective, informatisation is as much a system of power relationships and anonymous disciplinary strategies as it is a potentially emancipating agent.

The suggestion that the lack of information should be eliminated favours interventionist tendencies in public administration. Advertising campaigns in the computer and software industries also serve to strengthen the belief that more information leads to better decisions. At the same time, the sixties mythology of the democratic computer is still being propagated by businesses, cyberpunks and techno-anarchists. The Digital City is a good example.

Over and against the apologists of democratisation are the opponents of 'push button democracy' (Depla, 1995: 204). Informatisation can, it is true, considerably extend direct democracy but it also emphasises its 'staccato' character. Problems are simplified to binary proportions; opinion-forming accelerates to such a degree that cohesion and coherence evaporate.

It is interesting that this ambiguity is fundamental in the sense that contradictory tendencies usually make their appearance at the same time. Controls and services, centralised steering and autonomisation, democratising and blurring – all appear together in the process of informatising public administration. That is why it is naïve to adopt a voluntaristic perspective of ICT's significance. It is naïve because it is decisionistic and anthropocentric. It believes in achievability, whereas technological and social developments are, in fact, relatively accidental results of each other. That would seem to argue against a deterministic perspective, allowing us to close the discussion along constructionist lines. But I have objections to that as well: as a combined outcome, the developments are, in fact, determining their own future. And in that combination of ICT and social development, ICT is historically particular: it is a technology which de-centres the subject as the deciding instance. It decentres the subject as citizen, the subject as *citoyen/republican* and the subject as politician/sovereign. This radical outcome is technologically determined because without ICT it would be unthinkable and because ICT can neither be thought away nor eliminated. The political and social meaning of informatisation will be dealt with in the next chapter.

5 Politics

5.1 CRISIS AND CONTINUITY: INTRODUCTION

Complaints about politics and criticisms of democracy are nothing new. It is probably a sign of political vitality and the quality of the democratic system, that criticism is not only tolerated but is accepted as an intrinsic and authentic part of it. Diagnoses that suggest that the political-administrative system is in crisis and that large-scale transformations are in prospect should therefore be treated with some degree of scepticism. (See also Van Gunsteren & Andeweg, 1994.)

A number of initial observations and remarks need to be made:

- In the 1994 elections, the Christian Democrats and the Labour Party suffered unprecedented defeat. Furthermore, the electorate seemed to be completely adrift. About 50% were floating voters and 34 parliamentary seats changed hands.

    The first interpretation might be one of crisis. The divide between voters and politics is wider than ever before. Traditional parties have lost so much support that the system has become unstable. Extremist and single-issue parties are disrupting a system which relies on consensus and balance.

    A second interpretation is the diametrical opposite. Voters are no longer voting according to class, birth or belief. They are making rational choices as emancipated individuals, and one can now speak of a mature democracy. The high turnout shows that there is no divide separating voters and politics, but the high percentage of floating voters demonstrates that politicians and parties can no longer take their support for granted.

    A third interpretation could highlight increasing fragmentation and the decline of absolute values. Voters' choices are dictated by a wide range of criteria. Consistent over-arching value systems have lost their powers of persuasion. Politics remains important but its institutional forms no longer reflect social developments.

- In chapter two, I mapped the way in which different trends in organisation and governance are affecting public administration. New concepts of governance have come about and have been applied to the reordering and redeployment of various policy domains. Other patterns of organisation are modifying classical bureaucratic notions such as hierarchy and pyramid. Bureaucracy is being partly amended, partly differentiated and partly transformed.
  It is clear that these tendencies have not left political institutions unaffected. In particular, the fundamental primacy accorded to politics has been altered. Although, in some respects, it has intensified, on the whole it has been curtailed and reorientated. Our case study of independent agencies illustrated this. The many administrative and bureaucratic changes have also brought politics back into the centre of discussion.
- Finally, there are technological developments and their significance for public administration, which were discussed in chapters three and four. Various consequences have been observed, the most fundamental of which are: horizontalisation, autonomisation, deterritorialisation and virtualisation. In all cases, there are also consequences for politics and democracy, as was pointed out in the discussion of political primacy.

Nevertheless, there is no self-evident reason to assume that politics and democracy are in a state of crisis. Some might prefer to emphasise continuity: the democratic political system has given evidence of its resilience and vitality. Fukuyama (1989) even argues that liberal democracy has won such a comprehensive victory that history has come to an end. Others, like the De Koning Commission, point out that detecting political crisis is a common feature of politics at all times.

We shall argue that there is more to it than this and that our political system is under pressure from bureaucratic and administrative, technological and social developments. In the Netherlands this is reflected in the recent debates about administrative and constitutional renewal and about communitarism and republicanism. These debates are the subject of the following sections.

5.2   DEBATE AND SELF-REFLECTION

As we have said, a sense of crisis recurs regularly in respect of our politico-administrative system. Much less frequently, this leads parliament to explore the system's problems and to invite recommendations. Though that is putting it kindly. In recent years, the functioning of parliamentary democracy has been the subject of intensive debate in many circles but rarely among members of parliament. The setting up of the Deetman Commission in 1989 was a striking exception, and even more remarkable was that its report led to the setting up of four external and two internal committees to report back on its findings.

The setting up of these six committees gave the impression that parliament recognised the problems confronting our politico-administrative system and wanted to find solutions. There was even talk of 'institutional self-awareness' in which party political divisions weighed less heavily than the parliamentary system itself.

The Deetman Reports

In the report 'Effective Ministries', the Scheltema Commission deals with the subject of ministerial responsibility. In the wake of various scandals, followed in some cases by ministerial resignations, the report considers, in effect, whether ministers should be held responsible for everything their civil servants do. The significance of ministerial responsibility lies especially in the institutionalised relationship of control between parliament and the executive power.

The committee opposes any diminution of ministerial responsibility, a principle which is crucial in any democratic parliamentary system. But the responsibility should not extend beyond the sphere of competence of the minister concerned. Particularly where government organisations enjoy a high degree of independence, ministerial competence must be clearly defined and, preferably, applied uniformly across the independent agencies. In general, the committee recommends core departments, so that ministerial responsibility can be effectuated practically. Where the independent services carry out supervisory as well as executive tasks, ministerial responsibility will be more limited.

'Tuning the system', the report of the De Koning Commission, deals with the relationship between the electorate and their representatives. All but two of its members firmly reject the notion that the democratic system is in crisis. That notion, the Commission claims, is a historical and international constant of democracy itself. But empirical research, the report asserts, shows that, in fact, there is no question of a divide or of diminished confidence between voters and politicians.

The advice of the Commission therefore is to change as little as possible. In their recommendations, the system of election to the lower house is maintained; only the preference threshold is lowered. The powers of the prime minister are extended slightly. Election of the prime minister and the person charged with forming a government, is rejected by a majority on the Commission. Cabinet formation must be more tightly regulated. Election to the Upper House remains unchanged. It does get the right to send legislative proposals back, though the final decision rests with the Upper House. Finally, the Commission recommends that the political parties should look critically at the way they function.

In 'The Unchained Burgomaster', the Commission led by the former mayor of Amsterdam, Van Thijn, advises on whether mayors should be elected or appointed. Because of changes in local politics on the one hand, and the blandness and national dimension of local elections on the other, the Commission refers to a 'problem of modernisation'. The position of the mayor plays a role in this. This role would benefit from emphasising its co-ordinating and combining functions. A change in the appointment of the mayor might help. The aim is to strengthen local democracy and increase the effectiveness of local administration. The majority on the Commission support election of the mayor by, though not from, the council. Election by the voters is rejected (for the time being) because this might lead to a stalemate between council and mayor. Furthermore, the Commission does not want to add unnecessarily to the many other fundamental changes already taking place in local government.

In 'Towards Core Departments', the report of the Wiegel Commission, the structure and functions of central government are analysed. One of the main issues for the Commission is departmental regrouping. After extensive analysis of the current thinking about the organisation and functions of the Civil Service, which has appeared in a series of reports, reforms and operations during recent decades, the Commission outlines a number of basic principles. These consist of an acknowledgement of variety and multiformity, more attention to working procedures rather than to structures, better co-ordination of ongoing change, and an increase in flexibility. Departmental restructuring in that framework is unproductive and therefore undesirable.

The Commission considers at length the need for reorganisation of the Civil Service and recommends the establishment of core departments. These should confine themselves to 'strategic policy development' and 'political steering'. The implementation of policy should be passed to independent agencies. The Commission also presses for a general administrative service and a 'new-style project ministry'. Departments will continue to exist as units, but for important political matters a project-minister will be appointed to target subdivisions of different departments. Officials are appointed and placed with departments by the general administrative service (Frissen, 1993d & e).

The two other reports are internal. One deals with advisory organs and makes some radical proposals for reorganisation. The other report argues for greater decentralisation.

Caution and Imagination

Each of the reports tackles a similar set of problems: how to resolve bottlenecks in the existing politico-administrative system. These are the classic and modern problems of parliamentary democracy: classic problems such as the relationship between politicians (whether elected or not) and citizens, the legitimacy of a representative system, and the function of political parties; modern problems such as parliamentary control over complex and differentiated government bureaucracies, the structure and functions of the departments, and the relationship between central and self-governance. Many of these problems can be seen as problems of communication: between electors and elected, between parliament and government, between parliament and bureaucracy and (most common) between politics and society.

The reports make subtle judgements about the widely-assumed gulf between citizens and politics. The De Koning Commission, in particular, plays it down as primarily the concern of intellectuals who misunderstand the paradox of democracy, which necessarily creates higher expectations than it can satisfy. In any case, there is always a distance between electors and elected (Commissie-De Koning, 1993: 5).

This Commission's analysis is sometimes downright complacent. The crisis in the politico-administrative system, which others have noted, is now made to vanish with the help of a few diagrams on voting behaviour, attendance and party loyalty. The Commission seems to find comfort in the belief that the entire crisis was dreamt up by half-baked intellectuals. Only one member of the Commission (De Graaf), in a footnote as brief as it is venomous, distances himself from this analysis. Revealing is his observation that the Commission refused to chart the 'social developments of recent decades and the resultant changes in individual behaviour and the needs of the citizens' (Commissie-De Koning, 1993: 13). The majority of its members can not be absolved of a degree of pedantry.

In other reports, the tone is different: the Van Thijn Commission's analysis goes significantly deeper while some of the Wiegel Commission's proposals are quite radical. Many of the analyses and proposals, however, have an Archimedian perspective. The politico-administrative system is always the starting point of its reflections; social developments figure only as contingent factors.

There is no hint of a daring diagnosis or an imaginative proposal. In brief, the harvest of proposals consists of: 'reduction of the number of advisory bodies, scaling down of ministries concerned with policy, independent or privatised services for implementation, an administrative service at central level, greater decentralisation, indirect elections for mayors, and greater powers for the prime minister' (Korsten & Willems, 1993: 230). In other words, renovation, not change.

Most of the Commissions' proposals are directed to incremental change. Large-scale operations based on a coherent blueprint are rejected: they are costly, consume energy, are out of date as soon as they are implemented and more likely to cause things to seize up than to produce a cure. A more differentiated approach to problems is prescribed. Radical solutions are also avoided: mayors need not be elected, the voting system needs practically no adjustment, the person charged with forming a new government should not be elected, ministerial responsibility remains undiminished. Large-scale restructuring of departments and the introduction of a 'tower block' of directors-general (Wiegel Commission) is not recommended. And rightly so. Gradual, cautious and incremental adjustments are preferable to muscular and revolutionary élan. Grand schemes aiming at and promising radical change do not work. Furthermore, because pretentious, they are also dangerous. And yet. Is nothing then the matter? These reports leave one unsatisfied. There is something lacking.

Deficiencies

None of the reports touched on the heart of the problem, though the Wiegel Commission had the most to offer.

The key question is whether the position, function and significance of politics in our society is still the same as it was when our political institutions came into being. After all, they have hardly changed at all, whereas society has changed fundamentally. To put it another way, can a political system dating from the age of the steam engine satisfy the needs of the information society? The analyses and proposals of these Commissions should be judged in the light of that question. Current social developments have led, or will in future lead, to a society which is strongly differentiated and complex, with widely diverging patterns of values, norms and life-styles, and in which patterns of decision-making, and negotiation between divergent sectors have a networked character. From the perspective of politics and administration this means that the myth of manageability has to be significantly qualified. The desire to control society from a central point seems more and more to be an illusion – the self-steering capabilities of social connections will prevail. In which case, governance becomes, on the one hand, a relatively accidental result of a range of scattered developments, and on the other, a function of society as a whole and not of some specialist sub-system within it.

Our parliamentary-democratic system, however, is revealing quite different characteristics. In essence, the system is based on a pyramidal hierarchical conception of society. Parliament, at least in principle, is located at the top of that pyramid. It sanctions social consensus, it regulates social decision-making and, in broad lines, it makes decisions about the desired layout and direction of our society. Bureaucracies are the instruments which politicians control in order to execute parliamentary decisions. Such a political system seems anachronistic in the face of a modern society such as ours, and the results are there to be seen. Politics is being trivialised because ideologies are widely differentiated, and the grand integrating discourse can no longer be told, or is no longer listened to. Politics has become marginal because decision making takes place in networks of various actors who negotiate, compromise and fight their battles without any need for political decision-making. Politics has become dangerous because recent attempts to resolve great issues on the basis of a hierarchy and uniformity (health care and education) damaged social diversity and the capacity for self-steering.

The Commissions' reports appear to cling steadfastly to the existing system, their diagnosis is expressed in terms of failings, and the therapy in terms of gradual adjustment. The solutions come incrementally out of an idea of narrow margins. In all of them, politics as such remains sacrosanct. Politics is seen as the domain of fundamental balances and choices, without, it must be said, any reasoned argument being given. Politics, apparently, has that self-evident privilege but without any indication being given of what exactly politics is, or to what it owes that privilege. Furthermore, politics remains associated with, and sometimes completely confined to, traditional institutions. The fact that the public sphere in our society, including politics, has become more diffuse is not discussed. And in the end, during the parliamentary discussion of the reports, scarcely a single proposal was adopted.

The debate which actually does address the importance of the public sphere and the position and legitimacy of politics, is in fact going on elsewhere. It is the current debate about norms and values, about political and public morality.

5.3 RESTORED VALUES

The Deetman reports put into words some attempt at institutional self-reflection within a parliamentary democracy. The central question is about institutional significance: what values are represented by politics and, on the basis of these values, how is politics given form? A diagnosis of either content or form indicates poor communication. In the Deetman reports it is form in particular which is highlighted: what do communications look like now, what are the problems and how best can solutions for these problems be formulated? The actual content of the communication, between citizen and state, between the citizens themselves and within the state constitutes the topic of a different debate: the debate about a sense of community, public responsibility and citizenship. Common to both debates is the attempt to find answers to questions about the loss of legitimacy and functionality in the modern state. They are, as it were, the reverse side of the developments sketched in the preceding chapters. Whether adequate answers have been formulated remains to be seen. Let us first outline the debate.

Communitarism: an Outline

Communitarism or, to use a term frequently encountered in the Netherlands, 'community thinking' has both theoretical and practical roots. The most important theoretical framework is the critique of philosophical liberalism contained in the work of Rawls whose book *A Theory of Justice* has become, since its publication, 'the standard-bearer and hallmark of liberal political theory' (Van Seters, 1993: 2).

In liberalism, the right to individual liberty has primacy. It precedes and is therefore independent of what is 'good'. The consequences for the state are that it has no role in furthering 'the good'. The state has no conception of good; it certainly has no morality. Its task is limited to guaranteeing what is just, and justice is rooted in the freedoms of the individual. The state is neutral in both fact and principle.

The liberal conception of the individual is one in which freedom and autonomy are axiomatic. The individual is an independent and unitary entity, capable of applying norms and values, of setting goals, but which exists outside of them. Norms, values and goals are choices which may or may also *not* be made. Individuality is fundamental.

It is to this which communitarism raises objections. According to the communitarist way of thinking, the individual is, in fact, defined by norms, values and goals. By definition, the make-up of the individual is social, because the choice of norms, values and goals inevitably takes place within society. For communitarists, the conception of the person is therefore not individualistic as it is for liberals, but orientated to the community. Perhaps even stronger: a person is primarily social. The community is not some kind of voluntary association as liberals argue, but a matter of commitment and identity. The community is not external to, but inherent in each person (Van Seters, 1993: 2–7).

These essentially anti-liberal philosophical principles lie at the heart of a wide variety of theoretical movements in many different disciplines. All criticise the one-sided emphasis on the individual and his right to freedom, as philosophically limited and politically risky. Attention and respect for the community, which constitutes both individual and society, are essential.

Politics

It should be observed that community-thinking is primarily an American phenomenon whose most prominent representatives are Etzioni, Bellah, Glazer, Nussbaum and Selznick. This geographical imbalance is no accident. There may possibly be an element of distancing from the paradigms of positivism and behaviourism which dominate so many scientific disciplines. But it is more likely that communitarism is a politico-theoretical reaction against the 'casino capitalism' (Bovens, Trappenburg & Witteveen, 1994: 324) of the Reagan era. The decline of the public sector, the damage to social security and health care, the impoverishment of the urban ghettos – in short, the well-known 'American situation' – which resulted from this period are an important reason for the popularity of community thinking. It is no coincidence that many of Clinton's advisors come from these circles.

Community thinking, therefore, might be termed a typically American project: exciting and entirely relevant for American society, but much less relevant for Europe and the Netherlands. Here we surely do not need to rediscover the welfare state? Nevertheless, communitarism is generating a surprising of amount of support in the Netherlands.

Communitarism in the Netherlands

In recent years, communitarism has been attracting a growing amount of attention in scientific and political circles. This is because of the introduction and discussion of American writing on the one hand and on the other, through its adaptation and application by Dutch writers. American writing has received excellent and comprehensive treatment in a volume by Van Klink, Van Seters and Witteveen (1993). This work also provides a convenient survey of the debate and the position taken up by the major players.

Then we have the authentic Dutch contributions. Worthy of mention are De Beus (1993), Van Gunsteren (WRR, 1992), De Haan (1993), Hirsch Ballin (1993, 1994a), Klop (1993) and Van Stokkom (1992). Within that group there is a rough division between communitarism (De Beus, Hirsch Ballin and Klop) on the one hand, and republicanism (Van Gunsteren, De Haan and Van Stokkom) on the other. Communitarism in the Netherlands has been taken up mainly by Christian-Democrats and a section within the Social Democrats. Republicanism is encountered within Social Democracy, left liberalism and the ecological left.

As we have observed, community thinking is sharply anti-liberal. Liberalism, as a political movement, is attacked for its one-sided market thinking, laissez-faire attitudes and minimalist conception of the state. As an attitude to life, liberalism is associated with hedonism, calculism and a deficient sense of community. Listen to the distaste in the words of Hirsch Ballin:

> It is striking that we, as modern people, associate morality primarily with asceticism, with an aversion to sensuous pleasure, happiness and enjoyment. Norms and values are a nuisance, they are obstacles to a pleasant life and barricade the way to freedom (Hirsch Ballin, 1994a: 54).

Communitarism is critical of a range of symptoms and attitudes at various levels. There is concern about such social developments as criminality, fraud in welfare provision, mass culture and alienation. There is criticism of social permissiveness and public and political relativism. And there is resistance to a state which passively accepts consumerism, even in the social services, and adopts a position of moral and ethical neutrality. In this connection, Klop (1993) speaks of a politico-cultural paradox: while there is consensus on the undesirability of state influence on norms and values, at the same time there can be little doubt about its necessity and inevitability.

> The preferred resolution of the politico-cultural paradox (recognition of a legitimate cultural and political role, but restraint in its implementation and a wide sharing of responsibility PF) does not require the government to take the lead in teaching virtue, but does assume optimal morality in its relationship with the citizens. This extends beyond the minimal morality of the negative freedom of individual moralism, which is the hallmark of utilitarian liberalism (Klop, 1993: 212).

Without always arguing it specifically, it seems that communitarism, at least in the Netherlands, holds certain attitudes and life-styles responsible for the phenomena which it regards with concern and sometimes with repugnance. Liberalism as a political philosophy is identified with a libertarian public ethos, which has been on the increase in the Netherlands since the 1960s. And this far from consistent mixture of values, norms and opinions (a combination of provo, hippy and free marketeer) has led to a society dominated by consumerism, relativism and unreflecting toleration. This is then linked fairly directly to criminality, malingering, fraud and so on.

Set in a broader perspective, one sees the negative consequences of a modernisation process, legitimised by a utilitarian morality:

> Behind the social problems of our time there lurks a fatal connection between an ideal and an actual process. On the one hand there is the ideal process brought about by the Enlightenment, in which individual self-interest (as each of us chooses to perceive it) determines our behaviour – and of which utilitarian ethics is the outcome. On the other hand, there is the actual growth in scale and anonymity of social relationships in modern western society, whereby individuals, outside the private sphere, have to respond not to an 'other' but to an anonymous collective (Hirsch Ballin, 1994a: 71).

The most important consequence of modernisation is the replacement of substantive by functional rationality. This socio-cultural notion (see e.g. Zijderveld, 1983) is given a strongly critical interpretation within communitarism. Functional rationality leads to a 'technisation' and rationalisation of the world picture, and of the role of politics and the state. (See Klop, 1994: 116ff.) Instead of politics being intrinsically directed to validating norms and values, one gets essentially abstract, but procedurally and instrumentally specific, policy. A calculating, contractual society and a cost-conscious government are the actual consequences of a public and political ethos which is neutral and relativistic.

Communitarism sets against this the need for normative politics. Morality and ethics belong in the public domain, and politics can only lay claim to a special position if it creates the conditions for, and takes part authoritatively in a debate about the values and norms which concern the community. Politics has to be, and ought to be, about 'the good life' (De Beus, 1993). If politics abandons that pretension, one is left with nothing but a regulating mechanism of more or less adequate procedures and structures of public decision-making. Dutch communitarists point to 'wellsprings of meaning', as the traditions from which political debate about values and norms can and must draw. These wellsprings of meaning supply information about the foundations which underpin political positions: Jewish, Christian and humanist traditions.

Bovens, Trappenburg and Witteveen rightly make the point that the political community is always a derivation and can never itself be an authentic source for evaluating morality and ethics (Bovens, Trappenburg & Witteveen, 1994: 325). And it is this which, in their eyes, gives Dutch communitarism the characteristics of a political theology. Normative positions are underpinned by appealing to authentic cultural wellsprings which, in a way, gives rise to extra-political discussion. It becomes a search for truth, for justice and the good, not in a classical rational sense, but in a philosophical and theological sense.

That distinguishes the Dutch communitarists from the republicans, a movement with which the names of Van Gunsteren, De Haan and Van Stokkom are associated. They share the communitarist concern about the loss of a sense of community in our modern and partially privatised society. According to Van Stokkom (1992: 2–7), the responsibilities of the public are being visibly eroded by larger scales of operation, urbanisation and the expanding welfare state. Moreover, there is still much inequality, which threatens to become permanent, but little sign of action or organisation to alleviate it. For republicans it is citizenship, as opposed to the more communitarist sense of community, which forms the anchor point for a much-needed public reorientation. Citizenship implies the restoration of the *res publica,* while at the same time accepting the principles of plurality and respect for plurality.

> The common good which the citizen uses for reference is always a pluralist good: perceptions of the public interest will always be disputed. In principle, the public interest resides only where there is room for differences of opinion over what constitutes the interests of the community. What unites citizens is that they share and maintain public locations where differences can be articulated, argued over and resolved (Van Stokkom, 1992: 126–127; see also WRR, 1992).

This respect for plurality is very marked in Van Gunsteren's writing. In late-modern and post-modern society, plurality is a sociologically observable phenomenon which means that the community is unknowable – by either the citizen or the government. To contest this sociological reality is not just philosophically undesirable, it is unrealistic as a basis for social or political action. Because plurality is unavoidable, any modern conception of citizenship must be able to cope with plurality. To achieve this, it is not necessary to strive for a communal source of values, as the communitarists would like: 'mutually compatible behaviour will suffice' – nor is it even desirable 'it leads to a régime of schoolmasters, it seeks a metaphysical principle, it does not work and it is totalitarian' (Van Gunsteren, 1993: 6).

The citizenship that Van Gunsteren advocates is certainly not anti-liberal; individual autonomy is not only prized, it is a precondition. But at the same time, citizenship is the result of exercising that autonomy in dealings with other citizens. In this also lies the autonomous significance of public domain and political community. Any reference to metaphysical sources of meaning is superfluous and undesirable. Communities are recognised and protected, but so too is the right to dissidence and nonconformity. The republic is of importance, but there is certainly much more that needs particular attention (Van Gunsteren, 1993:5).

Van Gunsteren is a 'minimalist' republican and distances himself from the moralism of communitarists. That is less the case with De Haan and Van Stokkom. De Haan argues that republicanism not only concerns political decision-making, but also implies a judgement of the outcomes:

> For next to a model of deliberative procedures, it is the position of principle in respect of institutions which contributes to just and stable social relationships, and to the mental and social conditions in which participation in those institutions is made possible. This is an extension of the republican *credo* outlined earlier: the procedures of political life are an instrument for leading one's life, but political life can only fulfil that instrumental role if it forms a part of individual perceptions of what a valuable life can be. So it is not a choice between either procedure or substance; republicanism is both. (De Haan, 1993: 184)

Republicanism, too, embraces a public morality, even though it is more fond of plurality than communitarism. And in Van Stokkom we see that this public morality is directed not only at freedom and plurality, but also at equality. His position is somewhat ambiguous, in so far as he does not make clear whether he considers equality – especially material equality – to be merely a condition for participation in the political community, or whether equality constitutes a normative dimension of that community. In any case, it is obvious that Van Stokkom views equality as a condition for participation, and believes that participation in the political community is a civic duty which can lead to the integration of many groups (Van Stokkom, 1992: 93). And that closes the circle which joins republicans and communitarists: a good society is one with a moral foundation, whether it is based on communal wellsprings of meaning or the postulate of political participation as its norm and value.

Futile Romanticism I

Many objections can be raised against communitarism and republicanism, but they also contain many attractive elements. Juxtaposing criticisms and sympathy, however, would merely instigate a normative debate about their value as political theories or political utopias. Such debates are not very fruitful, not just because they are seldom resolved, but especially because empirically they are so problematic. After all, it is very tempting to ask communitarists and republicans, especially in the Netherlands, how adequate the empirical observations are, which in one way or another they need to shore up their arguments. The next step would be to evaluate the diagnosis based on these empirical observations. Only then would it make sense to pass judgement on the proposed strategies and utopias. Of course there is always a solution in 'political theology' as Bovens, Trappenburg & Witteveen (1994) have argued, but the debate soon grinds to a halt, however much one might enjoy the elegance and aesthetic of hermetic argument, as in Klop (1993).

If we look at the empirical observations of various communitarian works, a number of aspects may be noted:

In the first place there is frequent reference to growing inequalities and the rise of 'new poverty', usually considered to indicate a social divide. In terms of material prosperity, western society can probably not be properly compared to other parts of the world, but nevertheless there is evidence of increasing poverty, especially of groups who find themselves in situations of 'problem accumulation'. Inequality is revealed even more in unequal access to the labour market, social provision, education and the new media for information provision and communications.

In the second place growing insecurity is observed, brought about, on the one hand, by material and non-material inequalities, and on the other, by the sharp rise in criminality. This affects both the safety of public spaces, and the security of society itself, which is being destabilised by organised crime.

Thirdly, communitarists in particular often point to the weakening of relationships within modern society. This involves such widely differing phenomena as the breakdown of families, weakened social control and the disappearance of social relationships. In general, it is a process of de-institutionalisation leading to a weakening of social and cultural cohesion.

In the fourth place, both communitarists and republicans emphasise the sharp growth of dependency brought about by the welfare state. The expanded system of social provision, guaranteed and financed by the state, has adversely affected primary processes of care and help (see too De Swaan, 1976).

As they stand, these empirical observations are fairly unobjectionable, though one might point out that a distinction should be made between continental Europe on the one hand, and the United States and the United Kingdom on the other. The 'American situation' brought about by the wild capitalism of Reagan and Thatcher applies far less to Western Europe (Bovens, Trappenburg & Witteveen, 1994). At the same time, it should be observed that the problems of the welfare state in Western Europe are very different from the situation in the Anglo-Saxon countries.

But even with these nuances, the observations still make sense. More problematic is their diagnosis, which, to put it bluntly, is dismissive and critical. Criticism of material inequality does not have to be linked to a rejection of the de-institutionalising process. Moreover, the Western European liberal criticisms of the welfare state can not be compared to American communitarian pleas for a reformed welfare state. And finally, the disappearance of specific social relationships within a framework of classical institutions, does not necessarily point to the dissolution of *all* relationships.

Both the observation and the diagnosis seem to reflect a cultural standpoint which in a number of respects is anti-modernist. Social splintering, cultural relativism and economic transformation are indissolubly linked to modernisation. And it is precisely these processes on which communitarists train their weapons, some writers more effectively than others. Klop and Hirsch Ballin go the furthest and are quite explicit in their criticism of modernisation's suppression of substantive rationality, of the values and norms drawn from various wellsprings of meaning.

The restoration of norms and values, and a political community which encourages deliberation about desired norms and values, are the mottos which unite communitarists and republicans.

Of course to accuse these movements of moral regression would be an injustice. But there is a certain nostalgia in their work. And I certainly endorse the designation 'political theology' by which Bovens and his colleagues characterised communitarism. The return to cultural wellsprings may satisfy internal debate within certain political groupings, but for others it is merely obscurantist.

But also the anti-modernism revealed in the problem-diagnosis which typifies the cultural position of communitarists and republican should be rejected. At its heart there is a romantic attitude to life in which harmony and order are cherished values. By making social cohesion an absolute value, the eye is caught too easily by disharmony, chaos and anomie. At the same time, the restoration of values and norms can only be achieved by adopting community formation as a political and social strategy. It is true that there is respect for plurality, but even this is only valued to the extent that it is harmonious.

And so empirical observation and political philosophy bite each other in the tail. The circle is complete; the debate hermetic.

But their empirical observations can be given a different tone: one of fragmentation, relativism and individualisation. And depending on how one evaluates them, other political philosophies and strategies become conceivable. Ankersmit rightly emphasises the importance of the historically contingent, which much normative political theory seems to regard only as a reality to be combated:

> Ethical political theory reaches a judgement on the rights and duties of the citizen after the matrix of historicity and historical causality has been eliminated. In a word, ethical politics and political theory were, in essence, an attack on history, on historical causality and on everything which we can only understand historically. Historicity is arbitrariness, injustice, the deprivation of rights – a dimension from which a good society must be freed. (Ankersmit, 1994b: 135)

And although Ankersmit's criticisms are primarily directed at Rawls, I consider them to be directly applicable to many communitarists and republicans. Even if some of their observations can be accepted, it is right to be highly dubious about their political strategy of restoring values and norms to political and social communities. In the words of Bovens *et al.*:

> In short, developments in the infrastructure, such as globalisation of the economy, modern communications, high mobility of labour and capital, and the social dynamic and increase in scale resulting from them, are directly at odds with changes in norms and values desired by communitarists and republicans. (Bovens, Trappenburg & Witteveen, 1994: 331)

There in a nutshell is the theme for the next section: the shifting power relations between politics and bureaucracy within the apparatus of the state, and between the state and society.

5.4 BUREAUCRACY AND TECHNOCRACY: THE NEW POLITICS

Over and against the vain romanticism of communitarism and republicanism, stand developments in public administration. These developments are numerous and varied as we noted in chapter 2. The Deetman reports and the debate over communitarist thought and republicanism may, in more respects than one, be seen as a reaction to them. Firstly, they reveal an erosion of the legitimacy of public organisations, caused by these developments. Secondly, they provide a diagnosis of the problems faced by the political system. And thirdly, they contain elements of a revitalising strategy for the political system.

In earlier sections, I indicated what I believe to be the most important shortcomings of the Deetman reports, and why I do not believe communitarism or republicanism to be adequate responses. In the following section I shall summarise briefly the most relevant developments in the relationship between politics and bureaucracy before considering the specific political significance of these developments. There has been, in a sense, a shift, a displacement, in politics, as recently observed by Bovens *et al.* (1995). It is important to establish the institutional consequences of that shift and its implications for the phenomenon of politics.

Politics and Bureaucracy: Shifts

Between politics and bureaucracy there have been a number of important shifts, some more recent than others, which have been touched on in preceding sections. They fall into four categories: immanent tendencies, administrative innovation, technocratisation and informatisation.

Immanent tendencies
Earlier I cited Weber's perception of bureaucracy as characterised by great power, precision and persistence (p. 23 above). Marx too refers to it:

> Since by its *very nature* the bureaucracy is the 'state as formalism', it is this also as regards its *purpose*. The actual purpose of the state therefore appears to the bureaucracy as an objective *hostile* to the state. The spirit of the bureaucracy is the 'formal state spirit'. The bureaucracy therefore turns the 'formal state spirit' or *actual* spiritlessness of the state into a categorical imperative. The bureaucracy

takes itself to be the ultimate purpose of the state. Because the bureaucracy turns its 'formal' objectives into its content, it comes into conflict everywhere with 'real' objectives. It is therefore obliged to pass off the form for the content and the content for the form. State objectives are transformed into objectives of the department, and department objectives into objectives of the state. The bureaucracy is a circle from which no one can escape. Its hierarchy is a *hierarchy of knowledge*. (Karl Marx, 'Contribution to the Critique of Hegel's Philosophy of Law', in Karl Marx & Frederick Engels, *Collected Works*, vol. 3, (Lawrence & Wishart, London, 1975) p. 46.)

The phenomenon of bureaucracy is directed at domination, ideal-typically in the service of its master: politics. Politics supplies substantive rationality, which is then given direction by the functional rationality of bureaucracy. But in the process of modernisation there is an ongoing displacement of substantive rationality. Bureaucracy tends to push aside the evaluative and normative elements of politics. In essence, bureaucracy is in conflict with politics. (See Frissen, 1989: 75–81; Frissen, 1991: 17.)

Although this tendency is ideal-typical, one can also discover empirical evidence. The ascendancy of bureaucracy over politics is reflected in its sheer size and in the instruments at its disposal. Only the ethos of civil service loyalty prevents a gulf from growing between actuality and intention. Even so, it is only with difficulty that the hugely expanded bureaucracy of the welfare state can be kept or brought under political control.

While this tendency dates back to early times, and has already been described by Weber as an iron cage, it is being further strengthened by two more recent developments. They are, firstly, the fact that bureaucracy and social organisations have become increasingly interwoven, and, secondly, the phenomenon of 'bureaupolitics' (Rosenthal, 1988; Rosenthal, Geveke & 't Hart, 1994).

The interweaving of the state bureaucracy and social organisations, investigated by many in the context of network or configuration research (e.g. In 't Veld, Schaap, Termeer and Van Twist, 1991), causes a significant shift between politics and bureaucracy. On the one hand, it leads to a different structure in the policy cycle -more horizontal and interactive - through which politics loses the initiative in establishing goals. On the other, it makes it more difficult to alter retrospectively any consensus reached through such networks, leaving politics as little more than a rubber-stamp.

The phenomenon of bureaupolitics in the internal bureaucratic process has led to a blurring of the dichotomy between politics and administration. Bureaupolitics turns all the players into political actors, so that the primacy of political will no longer falls so obviously to politics. The bureaupolitical approach, however critically it is judged (see Rosenthal *et al.*, 1994), makes it clear that bureaucracy is neither a neutral nor a monolithic instrument in the hands of its political masters. Much government policy results from conflict between competing bureaucratic actors.

Administrative innovations

Much of what I described earlier as experiment, renewal and innovation, more generally referred to as administrative renewal, may be characterised as a bureaucratic reaction to processes of social differentiation and fragmentation: that is to say, the reaction of administrative and bureaucratic actors who want to adapt government behaviour to social complexity. The core of many initiatives lies within the bureaucracy, especially where bureaucracy connects with social actors, and where shared practices of intervention, steering and regulation are being developed (Kuypers, Foqué & Frissen, 1993: 47). Political influence on these processes is limited, if only because pushing back the prerogative of politics over substantive changes of course, usually marks the start of administrative renewal.

That pushing back, which is also partly a form of political reorientation, is apparent in the characteristics of administrative renewal outlined in §2.2. The reach of governance becomes restricted: less detailed, less intensive, more global, and reliant on social self-steering. The object of governance changes because input, output and interdependencies are chosen as connection points rather than the internal regulation of social sectors. The level of steering shifts to macro- and micro-levels. The involvement of social actors in governance increases. The perspective on steering becomes more bottom-up, and becomes orientated to pluriformity and differentiation as a starting-point rather than as a problem (Frissen, 1990).

Incidentally, this process of administrative renewal is evaluated in widely differing ways. Idenburg and Van der Loo represent a renewed appeal for the restoration of political primacy:

> Leaders are expected to set aside their technocratic masks and (again) develop inspiring visions of the future, political parties should again take charge of the public debate, parliamentarians must regain control over political decision-making, and The Hague, the political centre, should provide governance – though along general lines and in consultation with the citizens (Idenburg & Van der Loo, 1993: 135).

The leader pages of the quality newspapers frequently contain critical analyses of the technocratisation and bureaucratisation supposedly resulting from administrative renewal (Oerlemans, 1992; Von der Dunk, 1993; Ankersmit, 1994a). Others seem to regard the developments as inevitable, because of their inherent connection with social developments such as individualisation, fragmentation, informatisation and internationalisation (Van Dam, 1992; Bovens *et al.*, 1994). Meanwhile, politics itself is also holding its end up: criticisms of administrative renewal which lessens the position of parliament are being heard increasingly frequently. The 'purple cabinet' (a coalition of left and right) appears to be in favour of unravelling, clarity, uniformity and the restoration of political primacy. The question of independent agencies is an obvious case in point.

It is striking how often criticism of the technocratic and amoral consequences of administrative renewal are linked with communitarist or republican ideas. Restoring the primacy of politics requires the revitalisation of the moral duties of the state (Boutellier, 1994; Klop, 1993; Foqué in: Kuypers *et al.*, 1993;Van Stokkom, 1992; Hirsch Ballin, 1994a&b). And even a highly nuanced argument by Bovens *et al.* concludes with an appeal to the (moral) persuasiveness of the state and politics (Bovens *et al.*, 1994).

However, the debate does not change the fact that administrative renewal is primarily a bureaucratic initiative. The subsequent administrative arrangements communalise decision-making and lessen the primacy of politics.

Technocratisation and professionalisation
Technocratisation and professionalisation are inevitable consequences of extending bureaucracy in order to shape and perform a swelling number of state tasks. The state is intervening more frequently and more intensively in social processes which have become more complex. Entirely in conformity with the ethos of modernisation, it has led to a professionalisation of the bureaucracy. More and higher qualifications have been required, both individually and organisationally. The development of professional orientation and loyalty is therefore quite explicable. But that orientation and loyalty are in competition with the political code of meaning, which is sharpened by the growing professionalisation of politics. Advancing functional rationality as the core of the modernisation process leaves nothing and no-one untouched.

Increasing professionalisation within the state leads to a shift, or change, both in terms of 'discourse' as in terms of organisation. Professionalisation is, after all, an undirected as well as a directed process. As an undirected process, it leads to shifting power relationships between politics and bureaucracy because professional attitudes and practices become themselves self-evidently a 'discourse'. Political codes of meaning can then easily become regarded as 'unprofessional'. As a directed process (recruitment, organisational change or development), it is an expression of the 'discourse' and its surrounding code of meaning, while, on the other hand, it leads to a strengthening of professionalisation as an organisation process. In that process, power relations shift because in professional organisations, the taking of decisions is decentralised and local decision-making becomes more autonomous.

Technocratisation is closely related to all this. In the professional organisation, the power of the experts is great both by virtue of their prominent role in giving form to policy, and the technocratic ethos which prevails in such organisations. This technocratisation has been sharply analysed by Fischer (1990). He sees the process as a theoretically and technologically inspired means of coping with increased social complexity (Fischer, 1990: 57 ff.). Instrumental considerations of management and control lead to de-politicising of decision-making and deliberation. In his opinion, technocratisation is thereby a process of relative centralisation towards technical elites, and a reduction in social participation (Fischer, 1990: 179ff). That by no means implies that technocratisation is apolitical: there is a hidden agenda by which technical rationality as a superior code of meaning becomes the dominant decision-making orientation.

My own view is different, certainly in the light of what I said about administrative renewal. Although here too, there is a question of professionalisation and technocratisation, it is not as the primary goal of specific bureaucratic elites. In the process of administrative renewal, it is rather decentralisation and socialisation which occurs. However, that also fits in with the domination of practical, perhaps even of theoretical, administrative expertise. Decentralisation is, after all, a consequence of professionalisation of, and within, organisations. The socialisation of decision-making and deliberation is an administrative translation of insights into the network character of social and administrative reality. Decentralisation and socialisation can be seen as an expression of a new 'administrative technology' in which the design of effective decision-making is central. In the light of that technology, political preferences become the outcome rather than the starting-point of policy processes. By that I mean a growing orientation toward processes in the organisation of public decision-making, which lessens the primacy of politics (See Huigen, Frissen & Tops, 1993). That orientation is technocratic because it places administrative expertise at the forefront, and is to some degree neutral in respect of the political content. The great complexity of many large-scale projects has provided an important model.

Informatisation
As has been argued at length in previous chapters, the process of informatisation also contributes to the lessening of political primacy. Fischer sees a close connection between technocratisation and technology. In particular, because the dominant technology in the information society is knowledge-based, its advance in public administration makes (technical) expertise more prominent in policy processes and administrative organisation. The significance of this for the position of politics can be illustrated by a number of elements.

In the first place, informatisation leads to a shift in politico-bureaucratic relations because, as a set of tools, it is used primarily within the bureaucracy. The technological equipment of the Dutch parliament and of other representative institutions is minimal (Frissen, Koers & Snellen, 1992). Large systems and networks, and also the individual skills, are mainly to be found in the bureaucratic domain. Power over information and communication – traditionally, in any case, almost a bureaucratic monopoly – is becoming even more one-sided. And it is precisely that power which is so important in the political game.

In the second place, the development and management of information and communication technology have a logic of their own. Methodologically, the opportunity for political intervention in the process of system development is very limited (Van de Donk *et al.*, 1990; Schokker, 1994). In processes of synchronous development of systems and legislation, political intervention has had to submit to technically-induced discipline. In the relations between highly informatised executive organisations and departments, it comes out in the more institutionalised 'implementation test' to which political wish-lists with system technical implications are subjected.

Furthermore, the growing importance of communication technology makes it necessary, if only for considerations of security, to regulate authorisations and define them more closely. That too can curtail the political space.

In the third place, as the quality and capacity of ICT increases, decentralisation of administrative power occurs, and the 'intelligence' of local units in organisations and organisational contexts grows. This encourages the tendency to professionalisation and technocratisation referred to earlier. In their relationships with political decision-making, therefore, bureaucracies have become more powerful and more intelligent. Not only can they implement policy in a more decentralised and independent fashion, but they are also better able to develop policy themselves and influence the agenda.

In the fourth place, in the development of ICT itself, communication is becoming technologically and organisationally more important. As has been said earlier, that leads particularly to a process of horizontalisation within and between organisations. Horizontalisation means a relative reduction of hierarchical governance, making political intervention more difficult. Moreover, there are more actors involved in this technologically-equipped decision-making process, which again reduces the relative influence of the political actors. Zuurmond (1994) points out that horizontalisation, within an ICT infrastructure and architecture, can also be regarded as a form of discipline.

In the fifth place, there are de-territorialisation and virtualisation, outlined earlier, both of which can lead to a structural and substantive lessening of political primacy. On the one hand, as obstacles of time and space become less significant, the rationale for territorial division in the politico-administrative system is reduced. On the other hand, virtualisation may have far-reaching consequences for primarily political codes of meaning, as the power and quality of ICT-generated images and events increase.

Political significance

How should one now interpret the shifts which we have been considering?

First of all, there is the question of power in the sense of sovereignty or primacy. We have seen that from shifts in the relationship between politics and bureaucracy, bureaucratic power is increasing. There is a positional shift of power from politics towards bureaucracy brought about by immanent tendencies in their relationship. In a sense, it is like a tragedy: the more the political master provides his bureaucratic steward with the tools needed to perfect his stewardship, the more influential the steward becomes by virtue of his toolbox and the political dependence upon it. Administrative innovation, professionalisation and informatisation are also contributory factors.

Secondly, the developments which we have described seem to reposition the decision-making processes, on the one hand in terms of decentralisation within the political-administrative system, and on the other, in terms of extending participation. Political power is not only moved, it is shared between more actors. If we consider political power to be a relational phenomenon we see not only a shift from politics to bureaucracy in positional terms (one has more than the other) but also a shift within the bureaucracy (decentralisation as a result of professionalisation) and stronger links with social actors. That could be described as a restructuring of political power. More actors take part and the political content of the public domain becomes less one-sided. In administrative theory, this development is known as the multi-actor approach or the pluricentric perspective (see Teisman, 1992; Van Twist, 1995).

In the third place, they result in a modification of the idea that politics, bureaucracy and society form a pyramidal relationship, with politics as the principal and intrinsic centre. Public decision-making can better be conceived of as an archipelago of different centres. In other words, political power is less and less positional, and more and more relational in character. Politics, then, is the outcome of a process, and less reliant on specific positions or institutions.

Fourthly, there is a reorientation of the political-administrative system. A reorientation, it should be added, which is the result of developments described above and not, as Fischer (1990) appears to argue, the result of a conscious strategy or conspiracy. This reorientation affects the whole discourse of politics as a code of meaning, and pattern of behaviour. In that discourse, the administrative paradigm is increasingly predominant over the political paradigm. The administrative paradigm is professional and technocratic in nature and characterised by an orientation to technical and scientific rationality. Professional expertise in specific policy domains is important, but more important still is an orientation towards methods, procedures, structures and styles of decision-making in complex networks of political, bureaucratic and social actors.

Finally, this reorientation demands a complementary re-evaluation. Political prescription is less acknowledged as an *a priori* principle in processes of decision-making, assessment and deliberation. More and more, the emphasis is coming to lie on processes and style in public decision-making. On the one hand, the characteristics of process and style become more prominent as the object and point of application for steering; on the other, political norms relate to them more often. The procedural nature of decision-making is then the primary object to which political norms and values apply.

These aspects of political significance, which can be observed in the developments described earlier, form a number of patterns:

- Political power is shifting towards bureaucratic and social actors.
- The political arena is undergoing a reorganisation from pyramid to archipelago, from one centre to many centres.
- The political orientation is becoming more professional and technocratic: the administrative paradigm is becoming predominant.
- Political evaluation or prescription is being applied more to the procedural aspects of public decision-making.

From this can be concluded that displacement, reorganisation, reorientation and re-evaluation do not immediately justify the conclusion that politics is coming to an end. It is, at least, still possible to speak of a new politics – though in new places and in other forms. However, these places may be empty and the forms may be diffuse.

5.5 THE EMPTY PLACE OF POWER I

The foregoing discussion of developments which (can) diminish the primacy of politics reveals a striking combination of tradition, continuity, renewal and discontinuity. Although this was already pointed out in Chapter Two, it is worth repeating, refined by the observations of the previous section.

The picture of Dutch public administration, certainly in this century, has not been characterised by sharp demarcations between politics, bureaucracy and society. The dominant impression is rather that of institutional 'pillarisation' and (neo-)corporatism. Religious and social multiformity in the Netherlands has led to a politico-administrative system which in many respects is unique.

Firstly, it is characterised by the permanent necessity for compromise at the level of government formation. Stable majorities have required and still require coalition forming.

Secondly, the need for compromise has also been reflected in the personnel structure of the bureaucracy in which all the larger political movements are represented.

Thirdly, there has always been a strongly corporatist policy culture. Social actors played and play an important participatory role, both in deciding policy and in implementing it, through autonomously or jointly-run organisations.

Political power has thereby always been shared power: in coalitions, with social organisations; within that, the bureaucracy has been able to play a stabilising and mediating role. The primacy of politics has therefore always been relative: rather it was a case of the primacy of political and social elites who together governed the social base. Political scientists and historians have analysed this extensively (e.g. Daalder, 1990; Lijphart, 1968; Stuurman, 1983).

Against this background, one can speak of tradition and continuity. But there is, at the same time, renewal and discontinuity because the basis of the system has been dramatically changed by the diminishing relevance of religious and philosophical plurality. The substantive rationality of 'pillarisation' as a form of socio-political organisation has, in recent decades, been made obsolete by the powerful advance of secularisation. The forms of organisation still exist, but their ethos is much more one of professionalisation and modernisation, and the historical content of their foundations has become more abstract (Zijderveld, 1991). The organisational plurality is losing a corresponding substantive content. For this reason, Fortuyn (1992: 61ff.) has argued for a radical break with the surviving culture of consensus, compromise and coalition-forming.

But equally, there is continuity if we regard the developments described earlier with their political effects as the consequence of the modernisation process. Ongoing rationalisation of functionality affects politics. Bureaucracies are becoming increasingly important as are professional and technocratic orientations. A lessening of political primacy then results, while, paradoxically, the powers and responsibilities of the state are unprecedented. The growth of the state and its tasks seems, even at the political level, to be driven by intrinsic professional and technocratic considerations rather than by clearly delineated ideological programmes. One might add that, internationally, there is no traceable correlation between the scope of the state and the political colour of government coalitions.

The lessening of the primacy of politics in the modernisation process is one of the immanent tendencies in the development of the relationship between politics and bureaucracy. Disenchantment, the '*Entzaüberung der Welt*' affects politics and ideology.

The continuity of modernisation, however, also embraces renewal. This renewal is both instrumental and conceptual. Professionalisation and technocratisation in the politico-administrative system and its relations with social actors and developments receive expression in numerous instrumental innovations: different intervention mechanisms, forms of co-ordination, patterns of consultation, policy cultures. It is not for nothing that the theory of policy instruments is a prominent theme in public administration. The widespread interest in the instruments of policy and their provision has generated renewal, but at the same time that is an expression of ongoing modernisation: a penetrating functional rationality.

Conceptual renewal is mainly to be found in the development of new concepts of governance (Frissen, 1990). These concepts involve more than an alternative set of policy instruments, where effectiveness and efficiency are the primary motives. They are based not only on instrumental ambitions, though they are not unimportant, but on a new perspective on governance which includes both instruments and principles. On the one hand, they express the idea that alternative forms of steering are more effective (see Chapter 2); and on the other, they are normative, because they are partially based on a conception of desirable governance (see Bekkers & Van Donk, 1989). This desirability consists in: intrinsically modest steering ambitions and an orientation towards social participation. Its novelty is that a lessening of political primacy is seen not only as an inevitable result of social complexity and turbulence, but as an inherent component of the concept of governance.

These new concepts of governance and forms of administrative renewal reveal an important change: the orientation of politics and administration towards the process rather than the content of public decision-making and deliberation. To be sure, historically, that is a classic conception of democracy (see Foqué in Kuypers *et al.*, 1993), but in the development of the welfare state it constitutes a radical renewal, perhaps even a break. The emphasis on process – on structures and procedures (Snellen, 1987) – should be seen as a transition from direct steering by the state to meta-steering without overexplicit policy pretensions.

That is a clear break with the grandiose, all-embracing policy programmes of the welfare state. Policy – in any case often discredited by failures of implementation and lack of maintenance and maintainability – is moved more firmly towards the end of a process of policy- and decision-making. It has become the outcome not the starting-point, which demands, especially from political actors, an unusual degree of basic prudence and sometimes even some indifference. Political primacy is thereby reduced; where it can still make itself felt is in the validation of decision-making practices, and in the codification and sanctioning of their results.

The break, the discontinuity, which this reveals is primarily one of de-ideologising. Ideology becomes less significant as the starting-point and guideline for public decision-making. This is the result partly of autonomous politico-bureaucratic processes and partly also of more general social developments which will be considered in the following chapter. Politics does not thereby disappear, but the intrinsic orientation of political organisation changes. That change can be more fundamental than it appears at first sight if we relate it to the technological developments touched on earlier and their consequences of horizontalisation, deterritorialisation and relativisation of the subject. I shall return to this in later chapters.

The striking combination of continuity, tradition, renewal and break is resulting in new forms of politics and forms of new politics. Political domains and configurations are changing. They are becoming empty and diffuse.

What in my opinion we have been able to observe so far is a reconfirmation of the idea that the 'place of power' is empty. This idea which I have taken from Foqué can be traced back to writers such as Montesquieu, Tocqueville and Lefort (Foqué, 1992; Foqué in Kuypers *et al.*, 1993). The core of their conception of democracy is that a political philosophy which supplies a basis and norms for democracy should never be reduced to a 'voluntaristic conception of sovereignty' (Foqué in Kuypers *et al.*, 1993: 38). It is crucial that democracy creates a social awareness, and for that a capacity for self-reflection is necessary. The political involves both the content of that social awareness and the conditions for reflectivity. It has not always achieved this:

> State and government, which have modelled themselves on a voluntaristic and absolute doctrine of sovereignty, have been distanced from society and its growing differentiation. The isolation of an inaccessible, overburdened government is in danger of becoming even greater by a society which is itself crumbling and in danger of becoming amorphous under the pressure of a conception of man whereby the subject is driven only by a passion for equality and an ideal of negative freedom (Foqué in Kuypers *et al.*, 1993: 38–39).

Against that it is necessary to set a relational conception of political power. There, subjects are primarily actors linked with other actors in a range of interdependencies. Public space arises from their interaction and not as the creation of a state thought to be sovereign.

> It then appears, in the words of Claude Lefort, that the transcendental place from where a legal order can be designed and legitimated has essentially become an empty place. It is a place which no longer belongs to anyone and can not be mo-

nopolised by any specific ideology or world-view. In a democracy, the identity of a community is essentially intangible and latent (Foqué in Kuypers *et al.*, 1993: 39).

This does not therefore mean, according to Foqué, that the need for state interventions disappears. But it is important that those interventions should be seen and supervised, that they can be justified in terms of the general interest, and that they have no pretension to exclusiveness *vis-à-vis* the public interest.

The empty place of power is precisely the outcome of the developments which we have been considering. In one sense they empty the place of power by breaking through the ideological monopoly of political actors and institutions; on the other hand, they do so by pluralising participation in public space. It is interesting that this development can be seen as a consequence of a new administrative paradigm which diminishes political primacy and accepts professionalism and expertise as new orientations.

The great emphasis on the structures and procedures of decision-making and deliberation (meta-steering) is making politics more abstract and its content emptier. Moreover, as a result, the public space is expanding. Simultaneously, it is also fragmenting because actual decision-making is increasingly decentralised, local and functional. The demand for coherence is becoming more problematic. It can only be satisfied in retrospect and usually contingently. Invisibility and emptiness are consequences of a range of developments which are not intended, certainly not steered and whose final coherence is a coincidence.

In this, I depart from Foqué's conclusion. The empty space in my argument is not empty on grounds of principle, but is so through developments which lie enclosed in the modernisation process. This might mean that the fragmentation of politics is inevitable and that the primacy of politics will be increasingly difficult to concretise. To this I shall return in detail later. First, I wish to outline a number of political strategies which have arisen in reaction to the developments described so far.

5.6 POLITICAL STRATEGIES

Statements by politicians, their (rare) writings, and public debate show that the developments here described have not passed unobserved. In reaction to what is sometimes perceived as their threatening character, different political strategies can be formulated or, better, constructed – they are not always explicit. Four strategies can be distinguished: institutional fantasy, moral entrepreneurship, brokerage and restoration of primacy.

Institutional fantasy

The first political strategy is that of institutional fantasy. The term, borrowed from Depla and Tops (1992), indicates a variety of attempts to revitalise politics through changes in the relationship between parts of the political structure and through new forms of citizen involvement. It is a question of revitalisation because the place and significance of politics, as such, is not under discussion. In that sense, they fit into what Tops and Depla call the 'completion strategy' (Tops, 1994: 189ff; Depla, 1995). It is a case of system maintenance under pressure from political and social developments.

At the level of the relations between sections of the political structure we can mention as examples: the elected mayor, honorary chairmen of executive committees, the strengthening of the powers of the prime minister, the strengthening of the dualism and the role of representative bodies, the creation of urban regions or provinces and so on. At the level of citizen involvement in the political system, one might mention: the referendum, the opinion poll, city discussions, citizen consultations, quality panels, changes in the electoral system etc.

In this strategy, of which the Deetman reports are an important expression, there is a strong emphasis on intensifying political decision-making and deliberation, and strengthening representative democracy. It is recognised that legitimacy has to some extent been eroded, and an attempt to restore it is sought by strengthening the political dimension of the political system. That explains the strong emphasis on decision-making, control and responsibility in relationship-patterns. On the one side, there must be more opportunity for citizen participation; on the other, the position of the involved actors must be more clearly legitimated.

This does not amount to a redefinition of political and public decision-making. Its most important feature is a strengthening of democratic content. This raises the issue of strengthening of representative democracy, as well as of increasing the elements of direct democracy within a framework of representative democracy.

Moral entrepreneurship

The second strategy shows a strong kinship with the debates of communitarism and republicanism discussed earlier, and can clearly be recognised in the ideas and actions of Dutch Christian Democrats. A repeated and heart-felt plea to bring morality into politics recurs in their election manifestos, as well as in the writings of their intellectual representatives and some politicians (Hirsch Ballin, 1994b; Klop, 1993). This is very different from traditional moralising. For Christian Democracy it is a question of restoring values and norms to the public domain so as to strengthen moral reflection in public debate, and to base government actions on an awareness of community. Christian Democracy has its own views about the specific content of public morality, but, at the same time, it emphasises the relevant aspects of social and political plurality. It denies both the possibility and the desirability of politics which is not founded on morality.

However, communitarism is not exclusively Christian Democratic; similar ideas can be found among Social Democrats. But in Social Democracy, republicanism is much more prominent. Here, there is strong emphasis on the values and norms which (ought to) regulate the public space. The accent lies however on the task of politics as the protector of the public space and, within it, the necessary conditions for reflection and reflectivity. One condition is the possibility for citizens to participate in the community or the republic. To the extent that such possibilities are defined in material terms (see Van Stokkom, 1992, on minimum incomes) the dream of the 'good society' becomes more apparent, and the kinship with communitarism clearer.

The use of the term 'entrepreneurship' in naming this strategy makes clear that the moralising has consciously to be sought for and created. One can not just fall back on old forms and ideas as being self-evident. Given the recognised change in social relationships and given the observable modifications in culture and morality, a renewed definition of values and norms must be seen as an undertaking to which politicians and political parties can make an important contribution. In Wöltgens' words:

> Politics has, primarily, the function of directing society. This includes 'management by speech', or more broadly, the influencing of norms and values. Politics is not one-way traffic from autonomous citizen to the politicians; citizens form their views on the basis of all kinds of information from which they extract the arguments with which they support their opinions. Here is an opportunity for politicians. They can offer a framework within which a confusing stream of information can be made comprehensible and coherent; they can help the citizen to interpret the facts and provide a therapy for the diagnosis. In other words, politics can be the metaphor for an ideology. This also includes making (citizens) aware of new issues. Politics and politicians thus function as a 'moral resort'; they can create or destroy taboos (Wöltgens, 1992: 56).

And in a contribution by Bovens *et al.* (1994) which otherwise sees few opportunities left open to government and politics, we come across a plea for 'political vision and persuasiveness'.

Brokerage

This political strategy is in many respects the converse of the previous one. In this strategy, which I have called brokerage to highlight the important mediating role attributed to politics, there is a degree of political retreat. There is an acknowledgement that society can not easily be 'repaired' and that pretentious political ideologies have been discredited. Self-steering and self-organisation are not only recognised empirically, but viewed as inevitable. Politics and government action have to accept that as their point of departure, which is to say that self-steering and self-organisation should be seen neither as instruments within an otherwise directive framework, nor as problems to be combated.

The word brokerage symbolises the mediating orientation of this political strategy, which sees one of the most important functions of politics as the arrangement and mediation of social and bureaucratic processes. Politics is expected primarily to direct a play with various social and bureaucratic actors, without burdening them with *a priori* and explicit substantive preferences.

The point of departure for this political strategy is social variety. That means that a politico-administrative system must be developed in which the relative autonomy of social domains is made central. Within these social domains, all kinds of connections exist between society, bureaucracy and administration. The significance of this strategy lies in laying down connections between domains; guaranteeing the autonomy of these domains; and encouraging reflexivity between the domains (Teubner, 1983). The political *métier* thus consists of making connections where it is necessary, but especially to avoid doing so where it not necessary - herein lies the 'retreat' referred to earlier. Structuring and creating procedures for decision-making between and within domains become the central feature (Snellen, 1987).

The conception of democracy within this strategy is primarily horizontal in nature. It rests on a trust in self-regulation and self-organisation, and on the desire to encourage participation in decision-making. There is no universal prescription – because of social variety, only locally specific outcomes are possible. There is no 'grand narrative' about politics, and even less about the desired substantive outcome of local democratic deliberation.

In this strategy, the primary political function is mediatory: bringing parties together in an arena with a view to co-production of policy. (See also Teisman, 1992; Huigen, Frissen & Tops, 1993.) It is a director's function combined with strong intrinsic pragmatism.

In Dutch politics, it is a strategy which is found mainly within the Christian-Democrat and social-liberal political parties.

Restoring primacy

The final strategy has been gaining in popularity in the last few years. Its central platform is to strengthen or restore the primacy of politics in matters involving public decision-making. This strategy appears in a range of movements, considerations and operations:

- in discussions about the government's core-tasks, where an attempt is made to concentrate political decision-making on 'pure' or 'inalienable' tasks of state;
- in attempts to limit political steering to the main strategic lines of policy;
- in operations directed at the devolution of government tasks, in which political responsibility should be focused on policy development and determination;
- in criticisms of 'poaching' of public decision-making in the Netherlands, and resultant attempts to reorganise the consultation and advice structures in a range of policy domains;
- wherever a culture of efficiency emerges and where political leadership and initiative is more appreciated;
- in criticisms of the large and cumbersome defensive legal structures which surround large infrastructural projects.
-

Certainly, since the appointment of the 'purple cabinet' this strategy has gained in popularity and influence. Within Social Democracy, but especially among liberals, the primacy of politics is a 'great good'. They thereby distance themselves from the neo-corporatist decision-making culture, which during many decades of Christian-Democratic government has spread through so many policy domains.

The core of this strategy is to think in terms of clear responsibilities and a more dichotomous conception of the public and private sectors. Politics is identified with the public interest, to guard and formulate which is its primary duty. In practice, it leads to a preference for more sharply delineated responsibilities, the purging of public affairs and the disentangling of corporatist structures. Christian-Democratic principles such as subsidiarity and sovereignty of the social unit are rejected:

> The sovereignty of the unit arises from setting government alongside other organs, and only legitimising government intervention in situations involving the maintenance of law and order, the protection of the weak, or dereliction of duty. It further assumes that employers' organisations and trade unions derive their authority not from the government but from themselves. It fits awkwardly within a society

where democratic control, the general interest, and political primacy are of first importance (Bolkestein, 1992: 177).

Recent criticisms of independent agencies and privatisation – that they might go too far and undermine the primacy of politics – are expressions of this strategy.

5.7 AND YET ... THE GREAT SILENCE

It would be wrong to assume from the foregoing that within the political system there is a systematic and profound discussion going on about social and administrative developments, their consequences for politics and the political strategies required to fend off or absorb their effects. That is certainly not the case. Rather, there is an amalgam of reactions – sometimes ostrich-like behaviour, sometimes denial, sometimes hesitant self-reflection, and, all too often, silence.

It is true that the Deetman reports, and the political debates which they gave rise to, showed initially an awareness of the need to match many processes of administrative renewal with a complementary process of political renewal. But even in the analyses there has been predominantly a denial of the problems to which others had drawn attention, while many suggestions and solutions have died an early death. It is again striking that only the ideas of the Wiegel Commission were taken up in the subsequent politico-administrative exposé. And those ideas related primarily to bureaucratic questions. Technocracy is clearly very seductive.

The silence is even more deafening when it comes to technological developments and their consequences for politics and public administration. I have described those consequences extensively in the preceding chapters. It has become obvious that ICT is a strategic factor and that, as a result, political reflection on it is inevitable. There is, moreover, in public administration a steady and irreversible implementation of diverse ICT applications. But it is only the hype which arose in the United States about the digital highway which has penetrated the political debate and led, at the opening of parliament in 1994, to the announcement of a national programme of action, which was also published by the Ministry of Economic Affairs in 1995.

But virtually nothing has been said about the political implications of ICT. Where there is discussion, the instrumental approach is predominant and focuses on the application of technology to strengthen the existing system. Hence the prominent place accorded to combating fraud in the informatisation agenda of political-administrative circles. Election programmes too are primarily instrumental and interest in them is in any case minimal. (See Depla, 1995.)

The BIOS-3 memorandum (Ministry of Internal Affairs, 1995) signals a cautious change. In it, at least some attention is given to the contribution of ICT to democratisation and political renewal. But the contours of the existing political system, the system of representation and the primacy of politics remain undisturbed.

It is not clear how this silence should be interpreted. It is possible that my observations reflect expectations which are too high; perhaps the great silence is in fact a striking illustration of the social developments which I have been describing; perhaps it is another aspect of the lessening of political primacy which I have suggested.

Furthermore, many developments are contingent, certainly in the outcomes which collectively they generate, and therefore make it impossible to tell a consistent narrative which might lead to political self-reflection and, possibly, redefinition. Strategies are constantly 'emerging', as Mintzberg argues. Perhaps current political strategy and developments in the position and significance of the state can only be reconstructed in retrospect.

In any case, the politico-administrative and technological developments which have been described are neither isolated nor unique. They are taking place against the broader backdrop of social development. Economic, social, cultural and organisational patterns comparable to those described in previous chapters can be observed and to these we shall turn in the next chapter.

6 The Social Decor: Modernisation and Postmodernisation

Public administration, technology and politics each have a dynamic and a logic of their own, and in many respects are self-referencing. They are not, however, completely autonomous: they are constantly processing social and cultural developments. Of these, public administration and politics have a special position because they have society and culture as both their object and points of application for action and communication. Society and culture form the game, the frame and the language of politics and administration, as Van Twist (1995) would say.

Technology is special because, in many of its current variants, it has a bearing on information and communication, and thus touches public administration and politics at their core. And because information and communication technology is becoming increasingly intellectual, social and reflexive in character (Bell, 1979; Zuboff, 1988), our culture is becoming a technological culture (Dickson, 1990). There are innumerable interdependencies between the central developments of this book. These interdependencies can be seen as part of the process of modernisation, as defined in the sociological tradition. They also appear to produce patterns of fragmentation, as we have seen in previous sections. But fragmentation is a phenomenon which does not fit easily into the explanatory framework of the modernisation thesis. Many writers, therefore, prefer to draw the conclusion that fragmentation is evidence of postmodernisation.

This conclusion has been drawn from a sociological analysis of social phenomena. Philosophically and epistemologically, some have attached postmodernism as a consequence. That is a position which can be aesthetic in nature, as in architecture, literature, art, the cinema; or paradigmatic, as in philosophy and many scientific disciplines.

In this chapter, I shall first sketch a number of social developments to provide a decor of broad lines, striking aspects and the occasional detail. It is, of course, very brief. A few relevant fragments of the decor come up for discussion: economic transformation, organisational change and cultural patterns. I bring these fragments together in a few images with which I characterise present-day culture. These fragments and images together raise the question whether we are witnessing a process of advancing modernisation or the advent of postmodernisation. The observations and analyses of earlier chapters will then be briefly summarised. The thesis of postmodernisation will be developed theoretically in the next chapter. That theoretical elaboration can be seen as an intermezzo, as an overture to the following chapter in which I focus on postmodernisation in politics, administration and technology.

6.1 ECONOMIC TRANSFORMATIONS

Post-industrial Society

The development which present-day capitalism is going through is often described as a transition to post-industrial society (see Bell, 1973). In that society, the dominant domain in economic activity, both quantitatively and qualitatively – in the sense of a code of cultural meaning – should become the service sector. There are also plenty of statistics to substantiate this tendency (Harvey, 1989: 157). Accompanying this tendency is also the increasing prominence of information, and particularly knowledge, as the basis of production and service provision. The service economy is a set of relations of production in which information and knowledge are the primary strategic production factors and in which the access to information and knowledge brings competitive advantages. Automation has considerably increased the knowledge-intensity of production processes, while the process of information provision as such is taking on increasingly technological forms.

In this connection, Attali's observation is interesting: that the image of the service economy is too limited. He talks of a society 'which one could call "hyperindustrial" – a society in which services are converted into mass-produced articles of consumption' (Attali, 1992: 18). Here, the dual character of this development is indicated: on the one hand, the growing importance of information and knowledge in the industrial sectors; on the other, industrialisation of the service sectors. The connection between these two characteristics is informatisation. In terms of personnel, this development has led to a sharp expansion of the 'service class': those who, pre-eminently, are employed in the production and processing of information and knowledge (Crook *et al.*, 1992: 111ff.). Information and knowledge consequently become as important as capital and the means of production as a source of power.

Consumption and Consumerism

A second important characteristic of modern capitalism is the growing importance of consumption and consumerism. For the organisation of production, this means a switch in orientation towards markets and clients, and away from the production process itself. Moreover, this assumes the recognition of, and production for, sharply differentiated markets, with all the associated organisational consequences for businesses and organisations (Clegg, 1990: 209ff.). The result of this, and to an important degree its cause, is that status is being defined more and more in terms of consumption levels, a scale on which quantity rather than quality counts. Class relations as a derivative of production relations are losing their social and cultural significance (Crook *et al.*, 1992: 188ff.).

Mass consumption as the basis for capital accumulation is incidentally also the basis of Fordism (Harvey, 1989: 125ff.). The new element is that differentiation in mass consumption is growing. Consequently, economic development and its resultant socio-economic stratification will become less predictable, more fluid and more chaotic (Crook *et al.*, 1992: 124).

Internationalisation

The internationalisation of economic relationships is a third important development. Here too, ICT plays a central role, because it can sharply reduce the cost of innumerable transactions. 'Electronic Markets' arise (Malone *et al.*, 1987; Schmid, 1993) which facilitate information provision and communication to such a degree that obstacles of time and space are avoided. In this connection, Harvey speaks of a process of 'time–space compression':

> (...) the time horizons of both private and public decision-making have shrunk, while satellite communication and declining transport costs have made it increasingly possible to spread those decisions immediately over an ever wider and variegated space. (Harvey, 1989: 147)

It is important to understand that this process of internationalisation leads to wider and intensified global interweaving, and thus to important strategic interdependencies between economic actors. The operation of the financial system illustrates this clearly, and sometimes painfully (the crisis of Black Monday). Uniformity and globalisation are the outcome: the hamburger which is identical everywhere; the universalising of Coca-cola; the spread of TV shows; the dominance of Sega and Nintendo.

At the same time, the process of internationalisation includes apparently conflicting developments in the direction of differentiation and the small-scale. 'Time–space compression' turns the world into a village, and every village into the world. Economic activity can be organised very locally – this is how regional centres of economic activity arise. But also, thanks to ICT, international operations can be highly customer-oriented and market specific. Clegg (1990: 120ff.) uses Benetton as an example of a small-scale business operating on an international scale, because the large-scale connections lie 'hidden' within computer networks. Zuurmond (1994) also shows this, while Mowshowitz (1992) treats its effects on the state.

The Metaphor of the Market

A fourth important economic development might be called the 'metaphorisation' of the economic paradigm. Economic values and norms like the efficiency of market mechanisms and the importance of enterprise, are institutionalised and spread through economic and non-economic domains. Habermas might speak of 'colonising the life-world', but even without taking up a normative position in advance, one can observe a remarkable re-evaluation of the market economy's mechanisms and values. Thereby the market and enterprise become metaphors of efficiency and effectiveness which leave nothing undisturbed. This not only changes the classic dichotomy of state and market – even in the public sector, market and enterprise are models – but also puts the internal bureaucratic structures of businesses under pressure. Sweeping reorganisation leading to mass redundancies and permanent restructuring become legitimate because they remove inefficiency. Leaders acquire mythic proportions as new heroes of enterprise. I shall come back to the cultural significance of that later.

These developments – which are illustrative rather than exhaustive – mark a transformation, partly begun, partly emerging. That there are victims is evident, a fact which the management gurus who appear as apologists for this transformation seem, at times, to forget. But that the transformation is happening seems incontrovertible:

> (...) the (relatively) severe levels of unemployment experienced in some western societies towards the end of the twentieth century do not themselves represent the new form of society but mark the transition into it: just as the transition from feudalism/absolutism to capitalism (c. 1800–50) and the transition from liberal to organized capitalism (c. 1900–30) were marked by severe economic disruption at the personal level so also is the current transition. What is surprising is that the current transition is marked by so little (Crook *et al.*, 1992:134).

6.2 ORGANISATIONAL CHANGES

In the world of organisations and organising, we can also see a number of striking changes which are the result partly of the economic transformations which we have described and partly of the developments in the field of ICT. Many writers have pointed out the decline of the style of organisation, typified by Fordism and Taylorism, which focuses on division of labour and bureaucratic co-ordination. Its greatest drawbacks are rigidity and a lack of flexibility, which nullify the clear advantages of standardisation, routine, specialisation and manageability. Market or environmental turbulence demand greater adaptivity; complexity demands strong professionalisation; technological developments lead to more knowledge-intensity. Alternative organisational forms arise which have been given a variety of names: flexible specialisation (Harvey, 1989; Crook *et al.*, 1992); new production concepts (Fruytier, 1994); learning organisations (Swieringa & Wierdsma, 1990). In this connection, one might also mention orientations towards 'core business', 'just in time' production, 'lean and mean' production, quality management, business units, profit centres and outcome-responsible units.

To indicate the changes in organisations, I shall mention a number of relevant developments.

Flexible Specialisation

In the first place, a break with classical bureaucratic organisation is becoming apparent, both in the private and the public sector. The break is based on the increasing dysfunctionality of bureaucratic characteristics for organisations operating in a turbulent and complex environment. Those characteristics are: standardisation, formalisation, specialisation and centralisation. In conditions of stability and predictability, they can effectively bring about routine, reliability, security and manageability in business organisations. They become dysfunctional when those conditions alter under the influence of increased competition, internationalisation, informatisation and market differentiation. In such circumstances, response times drop, production cycles become shorter, the knowledge-intensity of production increases and markets become more specialised. That demands of organisations a different functionality: innovativeness, professionalisation and market-oriented work.

The other type of organisation which arises is characterised, according to various writers, by flexible specialisation (Clegg, 1990; Crook *et al.*, 1992; Harvey, 1989). Peters (1992) speaks of the necessity of 'disorganisation'. Flexible specialisation implies a permanent striving to keep production flexible and the pursuit of specialist niche-markets for specific products and services.

Organisationally this means: a wide delegation of authority, and granting as much autonomy as possible to organisational divisions; the professionalisation (through horizontal task integration and vertical de-specialisation) of the autonomous divisions; a more fluid organisational structure, which can be adapted to specific projects; a strong network orientation inside and outside the organisation; specialisation in specific products and services ('core business'); 'outsourcing' of non-strategic activities; a strong market- and client-orientation by improving services and customisation; the use of ICT as an alternative to bureaucratic mechanisms for co-ordination, information-provision and communication (see Zuurmond, 1994).

It is conspicuous that this form of specialisation, in respect of the internal organisation, makes de-specialisation necessary. Autonomous organisational divisions become responsible for integral business processes and their management. Within the divisions, professionalisation increases which is a spur to further autonomisation. But as Clegg (1990: 208ff.) rightly observes, flexible specialisation can equally well develop into an advanced control strategy within organisations.

Professionalisation

A second important development is the further professionalisation of organisations. There are a number of different causes and meanings.

In the first place, professionalisation in organisations increases because the position of the service and information sector is gaining considerably in importance for economic activity as a whole. We have already observed this. However, because of this, organisations throughout the economy are affected – through the buying in of services, the use of information, and communication with other organisations. That means that the management and steering of organisations are becoming more complex, and that the need for professionalisation is growing.

In the second place, widespread technological developments are making the knowledge intensity of production considerably greater. The substitution of unskilled labour by machines means that the total production process can be performed by fewer people. But they need to be better qualified. Moreover, as a consequence of the 'informating' capacity of ICT, the number of functions dealing with information provision and communication is growing rapidly (Zuboff, 1988).

In the third place, professionalisation is increasing as the outcome of a range of strategies in the areas of quality management and training. Better use of the intellectual capacities of 'human resources' in the organisation should thus contribute to the flexibility of production, to raising the level of service and to more effective market- or client-orientation (Swieringa & Wierdsma, 1990).

In the fourth place, professionalisation is on the increase as a result of decentralising decision-making powers to smaller organisational subdivisions. Consequently, more knowledge is needed at that level. At the same time as this 'push', there is also a 'pull': professionalised, decentralised subdivisions will use this knowledge to claim more autonomous decision-making powers.

Taken together, this leads to a situation where the professional organisation is not only taking an ever greater quantitative share of economic activity, but is also becoming a model for organisations in general. This image of Peters makes the point:

> Flat is a must, not an option. If communication has to go up and down the hierarchy, business is lost, and the firm gets an irredeemable reputation for unresponsiveness in an industry where responsiveness is almost everything. (...)Everyone has to be able to talk with, work with, everyone else, unimpeded – or else. The web of relations is the firm: survivors/stars are network virtuosos; losers/dropouts never come to appreciate the worth of major investment in network-building and relationship development. Forget the rules: 'My boss is my subordinate is my boss', depending upon the task and situation. So what? (Which is not to say it's easy to deal with. Neat freaks need not apply). (Peters, 1992: 181)

ICT

A third development which leads to fundamental change in organisations – and has already shown itself – is in the field of technology. ICT, in particular, is of importance because, on the one hand, it touches the heart of the organisation, and on the other, it steadily invades other technologies. I shall indicate the changes brought about by informatisation point by point:

- ICT strengthens tendencies to decentralisation in organisations because local capabilities of information provision and communication are expanding considerably. As a result, vigorous subdivisions can become ever smaller, which again tallies with the previously mentioned development toward flexible specialisation.
- A particular capability of ICT is what Zuboff (1988) calls 'informating'. ICT not only automates information provision and communication in organisations, but also generates different kinds of meta-information: information about business processes; information about the management of business processes; monitoring- and control-information; processed information about transactions. These streams of information enable the management to exercise more effective control, but also enable the creation and development of new insights in

respect of product development. Information, for whatever aims it may be collected, is becoming an independent, economic product, which in turn leads to input for new processes.
- The foregoing leads increasingly to organisational structures following the logic of information provision. That is the meaning of the phenomenon 'business process redesign' (see e.g. Davenport, 1993), a methodology which takes information relationships and communication patterns as its point of departure for designing organisations. A possible consequence is that organisational structure will increasingly become an information infrastructure and might thereby become relatively invisible (see Zuurmond, 1994).
- ICT makes obstacles of time and space less relevant. Immediacy can be achieved by on-line connections, proximity is possible without its being physical. On the one hand, this facilitates the internationalising of many organisations, and not necessarily involving anything large scale; on the other, it makes it possible to operate for a very specific market because all the necessary information can be concentrated in the relevant locations.
- ICT expands communication facilities considerably. Both networks and shared databases and applications make this possible. Furthermore, the widespread use of tracking devices (Snellen, 1994) plays a role. Organisations and clients can be linked together, without necessitating vertical integration within any single organisation, or the direct presence of clients in an organisation. The effect of this is that organisations become virtual: more communication system than physical artefact (Ettighoffer, 1993; Frissen, 1994a,b,c; Snellen, 1994).

Beyond Hierarchy

A final development related to change in organisations is, in fact, a combination of the three previous ones. Flexible specialisation, professionalisation and informatisation together lead to organisational forms which are distinguished by a greater degree of decentralised autonomy, more intensive horizontal connection and reduced vertical integration in the classic sense of bureaucratic hierarchy. There takes place a de-hierarchising of the organisation, which can no longer call on a '(...) centrally organized rational system of authority of which such spatial metaphors as 'hierarchy' can be placed' (Crook *et al.*, 1992: 187). What applies within organisations, also applies to inter-organisational relations. They too have a predominantly networked character and are regulated by contractual relationships of a more or less temporary nature. Mutual dependencies and strategic interaction jointly explain the decentred character of these inter-organisational networks. ICT facilitates such co-operative links to a high degree.

Autonomisation, horizontalisation and de-hierarchisation should not, however, be seen as utopian values. The organisational changes described here do not represent the 'good life' in a theoretical sense: power and inequality, conflicts and destruction remain. Their forms and images change. Their moral evaluation is another narrative, which will not be told here.

6.3 CULTURAL PATTERNS

Individualisation and 'Depillarisation'

There are many who have observed in our society a steadily advancing individualisation. (See also Frissen, 1994d.) Somewhat paradoxically, so Burgers (1989) argues, the processes of social narrowing, the consequences of technological, economic and spatial development, lead to a greater distance in social etiquette, and a crumbling of social connections. Following Elias, he points out that as the scale of society grows, and many kinds of relationships become more intense, so self-repression and self-control becomes greater. 'Secondary, distanced relationships become increasingly typical of social interaction' (Burgers, 1989: 315).

Economically, individualisation is a logical consequence of the market as exchange mechanism. Market relations are, in a sense, anonymous and assume that the actors behave as rational individuals whose choices are free of any social or cultural pressure. Spatially, individualisation promotes, in particular, the process of urbanisation. The city is distinguished by anonymity, and routine patterns of behaviour which enable large numbers of people to live together more or less harmoniously. Urbanisation and economic development have also considerably extended the variety of life-styles. On the one hand, this is the result of a continual division of labour; on the other, of a widening differentiation of consumption patterns. Socio-cultural identity is increasingly being derived from professional positions and from consumption patterns which are often deliberately conspicuous (Burgers, 1989: 316ff.). Greater variety is an important feature of these developments toward individualisation. A variety, moreover, which is not fixed: it is characterised by intensive and frequent change. As a result, group relationships, although they do not disappear, become much more changeable. Furthermore, the membership of groups becomes more changeable.

Institutionally, individualisation leads to the fragmentation of previously solid social connections. The family as the dominant social unit is in decline. Not only is the number of divorces high, so that in the course of their lives, people will have different family connections, sometimes simultaneously; but the number and variety of alternative forms is growing. Furthermore, an increasing number of people are living alone.

Traditional associations like churches and societies have also been heavily influenced. The church as an institution has lost much of its importance without this necessarily implying a decline in religiousness or spirituality. What has disappeared is church-going as a matter of course.

Clubs and societies have in no way declined in importance. But the traditional 'pillars' hardly play any role in providing cohesion. On the other hand, the importance of various 'single issue movements' has greatly increased. But relations between members is often limited to paying the subscription.

'Pillarised' society, so long the hallmark of the Netherlands, no longer exists. It is true that the organisational structure of the pillars has survived in many social domains, but they have little importance as meaningful or meaning-giving institutions. Individualisation is both a product and a cause of depillarisation. The disappearance of the pillar as a source of identity has greatly strengthened the elements of individual choice and individual autonomy. Relationships are no longer self-evidently linked to birth and family circumstances, but are objects of choice. At the same time, the greater importance of personal choice has removed much of the pillars' reason for existing. The social stability offered by pillarised society has thereby disappeared. The welfare state has taken over many of the pillars' functions. In that respect, the growth of the state has also contributed to individualisation.

Nevertheless, as different writers have observed, the concept of individualisation remains problematic. Wansink (1994) recognises the trend to individualisation, and observes that it leads to a growth of instability and of social and individual stress. As a consequence, there is a renewed desire for a social nexus, in particular the family, both for socio-economic and cultural reasons. Van der Loo (1994) speaks ironically of 'the new domesticity'. The desire for community is a reaction to cultural fragmentation and individualisation, though it is not inconsistent with it. According to him, communities do not come about on a self-evident cultural and historical basis, but are a choice, a question of management, even in the private domain.

I agree that there is a problem. From what has been said about individualisation, it is clear that it does not lead to the disappearance of the social nexus. In a number of respects, it leads to its growth. So in the world of politics and public administration, we observe more social groupings becoming involved in debates on a range of political and administrative questions. The environment is a good example. And in that same world of politics and public administration, there are now more social actors involved in the creation and implementation of policy than there were before in the limited participation-structures of the corporatist Netherlands.

Furthermore, the phenomenon of life-style, touched on above, indicates that the sense of identity offered by groupings in the private and public spheres is of great, and probably of increasing importance. Individualisation, though relevant, is inadequate as a metaphor for social developments. In my opinion, it would be better to speak of fragmentation as the dominant tendency in our society. I shall return to this.

Solidarity and Calculation

An important aspect of our society is the experience and design of solidarity. The welfare state is, after all, a specific historically evolved construction, in which the care for unforeseen situations and risks has been institutionalised. It usually involves the recognition of claims to a benefit or service which is not automatically available on the market. Solidarity always involves aspects of both altruism and calculation. Altruism expresses a concern for people and groups – a collective conception of civilised society. Calculation expresses, on the one hand, concern about future private discomfort, and on the other, concern for social stability.

Solidarity in our society is given expression at the level of specific collective services which, through taxation or insurance, we guarantee as 'uninsurable' risks. That involves all the services needed to guarantee a decent level of existence: benefits for disability, sickness or unemployment, a guaranteed minimum wage, basic training, help for the socially or culturally deprived and so on. As in many societies, the state plays a crucial role in this form of solidarity: it provides or organises the services or it guarantees them. De Swaan (1976) speaks of the 'statification of care arrangements'. Zijderveld too points to the statist nature of the organisation of solidarity (Zijderveld, 1983). This state involvement has, according to Zijderveld, also contributed to the commercialisation of solidarity in the welfare state. In socio-cultural terms, we could say that the substantive rationality of altruism has been pushed aside by the functional rationality of calculation. In that sense, one can also talk of a process of modernisation. Social differentiation has led to an extremely intricate system of provision in which, thanks to the available information technology, sub-systems have been created which take account of very specific circumstances and a wide range of detail. Solidarity is thereby turned into a question of calculation and, increasingly, recalculation. Intricate provision requires, after all, intricate controls, and technology offers some solace through the coupling of databases (see also Frissen, 1994b).

The ideological starting-point of solidarity is also becoming more abstract. There is still some talk of a global ideological consensus, but this is now under pressure. Financial circumstances are imposing retrenchment, while the specific content of solidarity, and the choices to be made, can hardly be argued ideologically any more. Social fragmentation is one of the causes. On the one hand, the solidarity-machinery is spinning out of control as every detail has to be recorded; on the other, as the number of details rises and their interrelationships change, any alterations in the system are out-dated at the moment of implementation. The social security system gives innumerable examples of this.

Ideological consensus, in the detail that would be necessary, is no longer possible. Added to which is that calculation is becoming culturally more important. Where exchange and contract are becoming more acceptable forms of social relationship, the cultural acceptance of all-embracing and organised solidarity is undermined. Any consensus then will, by definition, be minimal, and will relate mainly to a collective guarantee against uninsurable risks, which incidentally can only be established to a limited degree. Collective solidarity at the level with which we in our society are familiar, seems to be in conflict with the cultural, social and economic fragmentation characterising this society. In such a society, there is no longer question of a collective ideology which can support that level of solidarity. That is not to say that ideology is disappearing: the true significance of 'value relativism' is that there are many diverse and competing patterns of values. But because of it, collective consensus is necessarily minimal and, in its main lines, abstract in nature. Some cultural critics may regret it (e.g. Zijderveld, 1991); I merely wish to record it.

Mass Culture and Lifestyles

In the area of cultural orientations two important developments may be observed: on the one hand, the (international) growth of mass culture; on the other, a powerful proliferation of specific lifestyles.

Daniell Bell has already indicated the specific cultural development towards a mass culture, which is in conflict with traditional bourgeois culture in its links with capitalism:

> (...) bourgeois society of the nineteenth century was an integrated whole in which culture, character structure, and economy were infused by a single value system. This was the civilization of capitalism at its apogee. Ironically, all this was undermined by capitalism itself. Through mass production and mass consumption, it destroyed the protestant ethic by zealously promoting a hedonistic way of life. By the middle of the twentieth century capitalism sought to justify itself not by work or property, but by the status badges of material possessions and by the promotion of pleasure. The rising standard of living and relaxation of morals became ends in themselves as the definition of personal freedom. (Bell, 1973: 477)

In a later study, Bell has elaborated on his thesis of the immanent cultural contradictions in capitalism (Bell, 1976). An important consequence of this trend, which on the one hand, conflicts with the cultural consistency of capitalism (bourgeois culture), and on the other, is produced by processes of commodification and differentiation (Crook et al., 1992; 47ff.), is the disappearance of the normative distinction between the culture of the elite, (high culture) and the culture of the masses (popular culture). Intellectuals, in particular, resent it:

> This is perhaps the most distressing development of all from an academic standpoint, which has traditionally had a vested interest in preserving a realm of high or elite culture against the surrounding environment of philistinism, of schlock and kitsch, of TV series and Readers Digest culture, and in transmitting difficult and complex skills of reading, listening and seeing to its initiates. (Jameson, cited in Stauth & Turner, 1988: 519)

Greater geographic and social mobility and the advance of mass media and consumerism, create a mass culture which is, in essence, egalitarian (Stauth & Turner, 1988: 521). Cultural production in our society is placed at the same level as the mass production of consumer goods (Jameson, 1984b: 56). Crook and his colleagues (1992: 47ff.) speak of the commodification of culture. And that commodification involves, in particular, mass culture. The best examples can be seen in cultural productions on television and in the huge expansion of pop music.

The phenomenon of mass culture is very often seen as a form of international cultural homogenisation (Mommaas, 1993: 185ff.). McDonalds, Disneyland, Levi's and Coca-Cola are the icons which represent this cultural globalisation. It is very easy to talk in terms of cultural levelling and standardisation. Mommaas adds nuances to, and criticises, the homegenisation theory. In the first place, it is too static and ties culture too closely to the notion of territory. In the second place, it puts too much emphasis on the element of 'Americanisation', and misunderstands the context of what at first sight seem to be homogeneous cultural expressions. In the third place, the homogenisation thesis is 'étatist'. As a result, a spread of cultural elements through different national cultures appears to be homogenisation (Mommaas, 1993: 187–191).

The proliferation of lifestyles might be used against an interpretation of homogenisation. This too is linked to mass culture. For where the sharp distinction between high culture and popular culture is lost, so too is the code which establishes 'good taste' and 'correct style' (Mommaas, 1993: 181; Crook et al., 1992: 58ff.). Lifestyles are fragmented codes of behaviour from which one may, indeed must make a choice (Giddens, 1991: 81). Cultural identity is less and less automatically dependent on positional factors like class, sex, profession or religion. Furthermore, the exclusivity of 'high culture' as a model to be emulated is losing ground. This, too, is becoming another possible lifestyle which may or may not be chosen. The mass media and electronic communication offer an endless variety of cultural codes and styles.

The connection with class is also in a sense disappearing because style and taste are themselves becoming the primary parameters of social division (Crook *et al.*, 1992: 59). At the same time, variety is also on the increase. A good example is fashion. Where previously every season a specific image was decreed on the basis of *haute couture* (high culture) and then mass produced by the industry, in recent years fashions have become much more differentiated. Many styles exist side by side and outlive the fashion seasons, and this in turn has fundamental economic effects on the clothing industry. Equally powerful empirical evidence comes from MTV: every clip is a carefully designed icon of a specific lifestyle – in music, text, clothing, behaviour and image. Finally, there are children and young people, always a relevant source for cultural sociology. They automatically know and recognise lifestyles as cultural codes. Kuitenbrouwer (1990) has written a popular and instructive overview.

Cultural identity is becoming increasingly important. Of course, that has a lot to do with prosperity. However, we can also see that the significance of the cultural factor - in terms of style, fashion, design, image - is becoming more and more fundamental. Consumption is becoming central, so that production has to be geared to specific markets and niches. At the same time, the process of standardisation and flexible specialisation makes further differentiation of consumer goods possible. In this, cultural identities are both cause and consequence.

6.4 IMAGES

So far, I have given a fragmentary impression of the social decor surrounding developments in public administration, technology and politics. It has not been possible to talk of patterns of causation, only interdependencies. Nevertheless, to obtain a clearer understanding of social developments, I shall try to summarise them with a few free and improvised images, some of which I have borrowed from other writers.

The Contract Society

Economic developments are certainly not unambiguous. There is the transition to a service economy, but industry remains without doubt a strong base; consumption and consumerism are more and more the determinants of production of which differentiation is an important consequence, but also of importance is, naturally, highly standardised mass production; economic activities are becoming international, but regional economic centres are also becoming more important. What, however, links all these developments is their influence on social culture. Economics, as a code of meaning, is gaining power and is penetrating many spheres of life. There is, in other words, an 'economisation' of our world picture, which I referred to earlier as a 'metaphoring' of the economic paradigm. The discourse of the market and its juridical counterpart, the contract, is advancing and making its influence felt in the spheres of culture, politics, public administration and private life (see also Van Gunsteren, 1994: 221).

Fortuyn (1992) refers to this aptly as the 'contract society' in which the so-called 'calculating' citizen is active. That is in turn the result of greater prosperity and the success of emancipation movements and the welfare state.

> In the contract society, the emancipated calculating citizen must consistently be taken as the starting point for the governance of society. The citizen's freedom of choice comes first, in so far as it does not impede the choices of others (Fortuyn, 1992: 68).

This formulates in a normative sense the administrative and social consequences of the contract society.

But also in a direct economic sense, we can see the growing importance of contract as the pattern for connections between organisations and individuals. It displaces traditional ties like loyalty, consideration and organisation culture. Calculation, then, is another code of meaning (Van Gunsteren, 1994).

Archipelago

In the field of organisations, developments which are strongly influenced by ICT undermine the classic pyramidal concepts of organisations and organising. I have repeatedly used the archipelago as an image for this process (e.g. in Kuypers, Foqué & Frissen, 1993: 50). Others, such as Guéhenno (1994), employ the metaphor of the network to indicate a pattern of organisation in which horizontal connections within and between organisations eclipse the vertical relations, the pyramid, of hierarchy. Both metaphorically and empirically, ICT plays an important role in this:

> The naturalistic model, with its extensive structure of simple branches, gives way to the multi-dimensional model of the relational database. In a hierarchical, pyramidal structure, power is exercised by leadership and control. In the new structure, power is spread and divided over a large number of connections. Power is no longer defined in terms of control, but in terms of influence. This new form arises as a result

of the new communication techniques which facilitate the dissemination of information (Guéhenno, 1994: 67).

The image, therefore, is not utopian: power does not disappear; instead it is dispersed and acquires a strongly relational character. Flexible specialisation and professionalisation are important driving forces toward the archipelago form of organisation. The hierarchy as an organisational pattern of vertical integration, and the command-and-control model of steering grow obsolete. Again, that is no utopian thought, as some management gurus (Peters, 1992) would have us believe, but merely a description of developments which are not, in any way, bringing 'the good society' closer. It is much more a case of intelligent and perhaps technocratic adaptation, which causes mechanisms of control to change and in particular to become more anonymous and less tied to a subject or sub-system. On the other hand, De Geus points out that these developments in organisation patterns can easily be linked to libertarian political theories (De Geus, 1989: 224–227).

What applies to organisations, applies even more to society as a whole. It, too, reflects an archipelago or network character. Van Gunsteren and Van Ruyven (1993) speak of the unprecedented community, 'De Ongekende Samenleving' or DOS, giving us yet another computer metaphor.

Explanations for DOS are social plurality, problematic political representation, the social impact of expertise and government action (Van Gunsteren & Van Ruyven, 1993). The unprecedented community is, then, no longer steerable in the traditional way. What is left is steering through a pattern of variety and selection: the acceptance of variety and selective intervention (Van Gunsteren, 1994: 223). Whether that is realistic – especially since Van Gunsteren postulates the need for values and norms – is a question to which we shall return in later chapters.

Staccato Culture

The cultural patterns sketched earlier, in which the distinction between mass culture and high culture is disappearing and a striking proliferation of lifestyles is making an appearance, recur in the socio-cultural analyses and criticisms of Zijderveld. He criticises the advance of mass culture, especially at the points where it seems to take up high culture in its own banal discourse: it then becomes a matter of 'the amusement park syndrome', of 'blurred cultural boundaries (...) in a no man's land of generalised values and norms' (Zijderveld, 1991: 97-108).

Socio-culturally, Zijderveld considers the pattern to be readily explicable: in a process of modernisation, functional rationality increasingly displaces substantive rationality. And that leads to a situation where the values of functional rationality become, in a sense, new sources of meaning, while what remains of substantive rationality becomes more abstract.

At the same time, Zijderveld recognises greater individualisation and subjectivism, and increasing flexibility. The latter is naturally pre-eminently an outcome of functional-rational modernisation. The first arises as a result of the loss, or greater abstraction, of institutional codes of meaning. But the peaceful, clearly sign-posted transitions, which such codes claim to offer, also disappear. Cultural rhythms change from legato to staccato. A staccato culture is the result.

The heavy emphasis on image in modern culture is a further factor. The visual competes with the rational, the aesthetic with the philosophical, as Beunders has observed:

> All attempts to rein in the expansion of the visual media lose out to economic and technological development. Images fly through the air, crawl through the cable, become interactive and multiply with grim intensity and speed. (Beunders, 1994: 27)

The staccato culture is, to a high degree, an image culture. The image is bound up with a proliferation of lifestyles, whose external distinguishing features no longer represent deeper values and norms within, but are themselves those values and norms. Design is no longer something invisible which follows the logic of the function; design is an autonomous value. Hence, people can vary their lifestyles, sometimes at the same time, and lifestyle itself can be eclectic without being judged according to any particular rational consistency.

The picture produced by these images demands closer interpretation. Are we witnessing social development in the tradition of the modernisation process, or is that theory no longer adequate because we are dealing with a process of postmodernisation? In all probability, there is ambiguity: on the one hand, modernisation is continuing but in a more advanced form; on the other hand, there are qualitative discontinuities which point to postmodernisation.

6.5 ADVANCED MODERNISATION

Modernisation is a central theme in (classical) sociology. Marx, Weber, Durkheim, Simmel and other founding fathers used this concept to thematise and conceptualise the historical changes which they observed. Sociology can even be seen as an expression of the process of modernisation. And the theme has remained dominant in sociology. So much so, that the debate about it is conducted in eschatological terms. The end, or the crisis, of modernity is being announced: 'Modernity was a long march to prison' (Bauman, 1992: XVII).

And many define the present condition of western society as postmodern. Others insist that it is rather a situation of radical modernity, like Beck and Giddens who speak of 'reflexive modernisation' (Beck, Giddens & Lash, 1994). Yet others attach to these observations a normative defence of modernity against the relativising of postmodernism (Habermas, 1985a; 1985b; 1987).

To answer the question whether our society is in a process of modernisation, advanced or not, I shall adopt the conceptual framework of modernisation theory as applied by Crook *et al.*, (1992), Mommaas (1993) and Zijderveld (1983). My reasons for this extremely arbitrary choice are clarity and convenience.

Concepts

Crook *et al.* distinguish three central concepts in the process of modernisation: differentiation, commodification and rationalisation. They are mutually deeply interwoven, and all social phenomena and systems are determined by all three simultaneously.

Differentiation is a process of system specialisation by which systems can adapt better to external circumstances and demands while, at the same time, remaining internally coherent (Crook *et al.*, 1992: 7). In society as a whole, it leads to a continuous splitting up of domains through a multi-dimensional upgrading of specific functions (Crook *et al.*, 1992: 5). As a rule, differentiation is interpreted in a highly functional manner. Precisely because of that, its cultural assimilation is problematic, because cultural coherence, looked at socially, is also differentiated or splintered. At the same time, culture in general becomes more abstract and trivial, as we shall see with Zijderveld.

The process of social or structural differentiation has its economic parallel in the process of *commodification*. This concept is taken primarily from Marx and is defined as follows:

> (...) the process of commodification is one in which an increasing proportion of social objects are brought within the ambit of exchange relations, so that they are bought and sold for money in a market. (Crook *et al.*, 1992: 7)

The commodity has thereby become the fetish of modern culture. Primarily an economic category, it has penetrated every other domain including the personal. While many forms of regulation and cartelisation are modifying pure exchange relationships, the culture of commodification still persists and continues to develop (Crook *et al.*, 1992: 8).

The third relevant concept is *rationalisation*, which according to Weber is the most important engine of the modernisation process. Here it is primarily functional rationality: the rationality of ends-and-means thinking, of the procedural, of the instrumental. The process of rationalisation follows three interconnected lines. In the first place, scientific knowledge is given an ever more prominent role as the basis and legitimisation for action. At the same time the possibility of attributing meaning to action is reduced. In the second place, impersonality increases. Change and development break loose from tradition and personal expression. Economically, an invisible hand rules; politically, neutral regulations and anonymous bureaucrats rule. In the third place, rationalisation implies an impressive expansion of control and domination. Social relationships become the object of technological domination; their scientific, economic and administrative characteristics are made central (Crook *et al.*, 1992: 8–9).

Mommaas (1993) places *domination*, which he sees primarily as cybernetic control, in the centre of the process of modernisation. Following Parsons, he typifies the modernisation process as a process of differentiation; in particular, a differentiation of the social subsystem in respect of the cultural subsystem. A precondition is religious secularisation which, on the one hand, leads to pluriform religion and, on the other, to a generalisation of social culture. A further differentiation occurs between the economic and social structures as a result of the industrial revolution, and between the political and social structures as a result of the democratic revolution. The post-war educational revolution completes the modernisation process, for the time being (Mommaas, 1993: 32–35).

In more analytic terms, the process of modernisation proceeds

> (...) via four processes of structural transformation, namely: 1. improvement in adaptive powers, 2. increasing differentiation, 3. increasing integration of parts of the environment and 4. a generalising of values (Mommaas, 1993: 37, taken from Parsons).

Finally, Zijderveld's characterisation of modernisation is two-fold: on the one hand, there is *specialisation* or *structural differentiation*; on the other, there is *generalisation of values* or *rationalisation*. At the level of behaviour, there is a complementary process of *civilisation*, which should be understood primarily as an increase in subjective control (Zijderveld, 1993: 79–80).

Through structural differentiation, functional rationality will increasingly dominate substantive rationality, with all the problems of legitimacy that this creates. Institutions become functionally more specialist, while their cultural character becomes more vague and generalised. Because this applies throughout society, social culture also becomes more general and abstract. However, because the legitimacy of modern institutions is no longer substantively self-evident, they are developing a tendency to legitimate themselves, to justify their existence, in increasingly functional terms. They emphasise their organisational dimension, and thereby bureaucratise. Bureaucratisation is thus one of the most important dimensions of the modernisation process (Zijderveld, 1993: 83–100).

This brief sketch of the concepts in modernisation theory shows that in judging social developments according to their degree of modernisation, the following aspects are at least relevant: differentiation, commodification, rationalisation, generalisation, control and bureaucratisation. I shall use these concepts in my treatment of the fragmented social development, touched on earlier.

Economic transformations

The development towards a post-industrial society, as described in Bell (1973), is a form of further differentiation. In the totality of activity, the service sector specialises and also differentiates itself. A highly splintered pattern of services develops, from the poorly-qualified to the highly-professionalised. More recent tendencies in the field of information-services and the communications sector accentuate this pattern. Furthermore, there is no longer any question of service provision being secondary in respect of primary economic sectors such as industry and agriculture.

Differentiation is strongly stimulated, as well as caused, by the central position of consumption and consumerism in the capitalist economy. Markets are divided up into specialised segments, and production is structured to serve these segments as flexibly as possible and to produce further segmentation.

At the same time, the economy is being internationalised on a global scale. Thereby, the capitalist economy is penetrating every geographic region, taking with it the capitalist ethos: the metaphor of the market. To be sure, it takes the form of continuing economic differentiation on a world scale, with industrial decentralisation as an important outcome (Crook *et al.*, 1992: 29), but at the same time, it can also be seen as a form of increasing integration. This integration involves more particularly the totality of capitalist production relations. At the level of the individual business, the process is rather one of differentiation.

Commodification is another important aspect. I have already quoted Attali to the effect that, in the service sector, one can speak of 'hyperindustrialisation'. More and more services are being mass-produced. Once again the sector of information and communication provides a good example. Furthermore, the domains of cultural production and private life are falling under the effects of commodification through the entertainment industry.

Rationalisation is the driving force behind these developments. The knowledge-intensity in the forces of production and production relations is growing ever greater through technological development. At the same time, knowledge, information and communication are more and more becoming autonomous production forces: in many sectors, they are the determinants of competitive advantage.

But knowledge, information and communication are also important conditions for controlling the processes of economic differentiation. Economic activities on a global scale, organised along decentralised and differentiated lines, can only be controlled by means of ICT. Whether the degree of control is increased by it is uncertain. Some writers, like Offe (1985) and Lash and Urry (1987) flag up the increasing disorganisation of capitalism. The bureaucratic model of control, and state intervention, are losing their effectiveness. Others such as Zuurmond (1994) argue that technology has taken over this control.

Changes in Organisations

The organisation has always been seen, and certainly in its bureaucratic guise, as a phenomenon pre-eminently associated with modernisation. Organisations are artefacts of the modernist urge to control; they embody functional rationality. Bureaucracy is an instrument in the hands of its masters with which they attempt to control a specific domain or territory – both legally and by formalised and standardised practices. During their history, organisations have been the object and product of ongoing differentiation. Important elements of organisation theory deal with reception, and incorporate differentiation explicitly as an instrument of management. Contingency theory and the work of Mintzberg in particular, are attempts to establish satisfactory levels of differentiation in varying environments.

Economic and technological developments constitute the driving forces behind this differentiation. Further segmentation of markets, globalising of economic relationships and intensified competition make permanent differentiation a necessity for survival. At the same time, there is autonomous technological development – both in the area of production and that of information and communication – which promotes and facilitates differentiation.

As we have seen, however, economic developments and transformations in the field of ICT lead to organisational changes which break through the classic – in the sense of modern – patterns of bureaucracy. Flexible specialisation is just such a change, which can signify an amendment, a differentiation, or a transformation of the bureaucracy. It is an amendment in so far as it removes dysfunctionalities in the bureaucracy by a process of continuing functional rationalisation. It is differentiation in so far as decision-making powers are coupled increasingly with the immediate production processes. But flexible specialisation is equally a form of transformation in that it breaks through the rigid hierarchy of bureaucracy and leads to de-specialisation at the level of semi-autonomous units, and breaks with the linear process of differentiation.

It remains, of course, possible and realistic to see these forms of organisational change as examples of the pursuit of dominance. The literature on this still has a strong bias towards management. Continuing commodification, stimulated by consumerism, is an important driving force.

Nevertheless, some qualification is necessary. The process of professionalisation, taking place within organisations as a result of increasing knowledge-intensity in production processes and increasing knowledge content of the product itself, leads necessarily to greater autonomy for the 'shop-floor', decentralisation of decision-making powers, and decreasing importance of hierarchy as the co-ordinating mechanism.

Such processes of professionalisation also change the patterns of meaning in organisations: these too become decentralised. The codes become local in character. At the level of the organisation as a whole, we now only see the arbitrary and trivial meanings contained in 'mission statements'. These are practically identical in all organisations, which can be explained by what Zijderveld calls a process of generalisation as culture becomes more abstract. It is therefore not surprising that organisation cultures are so popular: they can be seen as a forced attempt to instrumentalise meaning within organisations in the interests of control. (See Deal & Kennedy, 1982: 193ff.; Jorritsma-Mientjes & Frissen, 1988.)

In conclusion, it can be stated that, at the level of the organisation, ambiguity predominates. On the one hand, we see the dismantling of bureaucracy and hierarchy; on the other, we see functional rationalising and the pursuit of control. We see continuing differentiation, but also generalisation and de-specialisation. It is still possible to interpret change as instrumental in respect of the dominant patterns of the modernisation process. Vicious circles in bureaucracy are simply broken in order to rationalise organisations. But at the same time, there are also breaks and discontinuities when we review flexible specialisation, professionalisation and specific characteristics of ICT.

Cultural Patterns

Individualisation and depillarisation are clear forms of differentiation, even though they result from processes of condensation. Economic rationalisation assumes that the actors behave and calculate as individuals, either as producers or, especially, as consumers. Patterns of consumption are becoming more and more the basis of cultural identity. Professional positions, too, are a part of the proliferation of identities, which is a logical outcome of the division of labour (Burgers, 1989: 316ff.).

Rationalisation is therefore the economic basis leading to commodification and differentiation. However, we can see at the same time that in a cultural sense, the pattern of differentiation is less rational and straightforward: variety without fixed forms, and fragmentation which is volatile and accidental. I shall return to it later.

Depillarisation can be seen as a process of de-institutionalisation: functional rationalisation undermines the cultural content and specificity of social ties, and generalises them. The pillars are now merely organisational structures; their culture has become abstract. But there are also alternative patterns of culture formation: more variety in living styles; religiousness and spirituality without the traditional church-going; associations and 'single issue movements'. It is true that the values and norms which the pillars represented have been generalised, but that does not justify Zijderveld's conclusion that in a cultural sense, society as a whole has become more abstract. Economic and social differentiation and rationalisation appear to lead to cultural fragmentation. That fragmentation is in its turn a spur to economic and social differentiation, boosted by the process of commodification. The popularity of communitarism (see §5.3) might well be explained as a regressive reaction to these patterns.

Individualisation and depillarisation have contributed to the further rationalisation of the organisation of solidarity within our society. The nationalisation of welfare arrangements is the functional substitution of the primary ties of solidarity. An ideological consensus at the level of the state shores up this functional design. But differentiation of specific arrangements and the erosion of ideological consensus, also as a result of an advancing culture of calculation, have led to cultural legitimacy becoming so abstract that it is no longer adequate. In giving form to solidarity, individual calculation and individual liability are acquiring more and more the status of cultural values. Collective ideological consensus can no longer be formulated, because the process of differentiation is moving too quickly. Abstractions can no longer provide the required legitimacy. At the same time, cultural fragmentation is such that fixed arrangements, however differentiated they may be, are obsolete at the time of implementation. The individualised culture of the calculating citizen is as much a result of this development as a more appropriate form of legitimation.

This culture of the calculating citizen is the cultural parallel to what I called earlier the metaphorisation of the market. It is the economisation of value orientations, and as such is a logical consequence of commodification and rationalisation. On the one hand, it is process of cultural generalisation; on the other, it strengthens economic and social differentiation. On balance, it leads back to cultural fragmentation.

This cultural fragmentation can be seen even more clearly in the proliferation of lifestyles. This is the consequence of the gradual disappearance of normative opposition between high culture and mass, popular culture. The interface with the calculation culture is the necessity of choosing between lifestyles as fragments of mass culture (Giddens, 1991: 81). Designing one's own life is a choice which can not be avoided, and is stimulated strongly by the image culture of consumerism.

Lifestyles do not have a hierarchical order: good taste or correct style can not be defined; even the notion of 'avant-garde' is popularised when renowned architects are not ashamed to design the ultimate symbols of mass culture such as Disneyland.

Differentiation, rationalisation, control and generalisation are becoming less appropriate as explanatory concepts for this cultural development. Mass culture and lifestyle, in particular, are transient, accidental or 'ephemeral' (Harvey, 1989), and not the linear consequences of a process of differentiation which reflects a functional, normative or in particular a hierarchical pattern. It is a process of fragmentation that has advanced powerfully through economic commodification and now looks to become its most important force. In the words of Crook *et al.*:

> (...) the progressive differentiation of culture, society and personality characteristic of modernity involutes so that the very idea of an independent, purely social structural realm no longer makes sense. Rather, 'society' must be understood in terms of 'culture' as patterns of signs and symbols penetrate and erode structural boundaries. (Crook *et al.*, 1992: 35)

6.6 NUANCES AND DISCONTINUITIES

The Decor: Economy, Organisations, Culture

The question whether these fragmenting social developments indicate a fundamental break with modernity requires a qualified response. If, with Giddens and many others, capitalism and industry are seen as institutional dimensions of modernity (Giddens, 1990: 55ff.), then it is merely a matter of gradual change, in so far as industry has been displaced by service provision and knowledge production as the dominant patterns of economic production. Technological developments are an important contributory factor. Furthermore, as capitalism is globalised, so too are the institutional dimensions of modernity (Giddens, 1990: 71). Seen in this light, capitalism, as the economic shape of modernity, has, since the disappearance of competing economic systems, enjoyed thus far an unimaginable hegemony.

At the same time, consumption seems, more and more, to be becoming the driving force of the capitalist economy. According to Bauman (1992: 50–51), that makes a society postmodern. Freedom *and* control are guaranteed simultaneously: the consumer society is at once liberal and panoptic (Bauman, 1992: 97ff.). Seduction and repression go hand in hand. And Lash insists that in economic terms it is becoming increasingly difficult to distinguish between production, distribution and consumption (Lash, 1990: 11ff.). The economy seems to be becoming cultural: meanings and symbols, though not structured hierarchically, move to the centre. Advanced capitalism has become an economy of meanings. The image, or Baudrillard's 'simulacrum' (1985) becomes, according to Jameson the final form of 'commodity reification' (Jameson, 1984b: 66). In his eyes, postmodernism is then nothing else than the cultural logic of late-capitalism. We have encountered a similar analysis in Bell (1973, 1976). Economising the world picture, the metaphor of the market, efficiency and calculation as new cultural norms are all phenomena which bear witness to the symbolisation of economics and the commodification of the symbolic.

So we have nuances, subtle modifications, and we have discontinuity, a break. On the one hand, there can be no doubt that capitalism is the dominant economic system, however 'advanced' it may be described. In that sense, one can only speak of postmodernisation in a highly qualified manner: the dialectic of two competing systems has disappeared and is replaced by a plurality of more or less late-capitalist economic relationship patterns. On the other hand, the driving force of the economy is increasingly cultural in nature: consumption and its associated mass culture and variety of lifestyles. The dominant production forces at this moment – information and communication technology – are stimulating this 'culturalising' of the economy with their emphasis on images, meanings and knowledge, and doing it on a global scale. (See Giddens, 1990: 77; and also Jameson, 1984b; Guéhenno, 1994; Crook *et al.*, 1992; Harvey, 1989). Culture becomes in a sense a production force, but it is culture primarily in a varied, plural and fragmented sense. That constitutes a break with 'classical' modernisation.

Also in the area of organising and organisations, we can observe nuances and discontinuities. It would be naïve to assume that bureaucracies are disappearing, that all organisations are ceasing to be hierarchical, and that management and leadership are becoming obsolete qualities. Management gurus like Peters (1992) with their infectious verbal violence often appear to be giving that impression. The irony is that they tell this story, for a not inconsiderable remuneration, to audiences of managers, who allow their autocratic positions to be criticised as anachronistic. Nevertheless, organisations are changing. The rigidity of the bureaucratic model is inadequate in extremely turbulent market relationships, which are not only distinguished by heightened competition, but also by a high rate of change and a proliferation of consumer needs. Flexible specialisation is one answer (Clegg, 1990; Harvey, 1989). Flexibility and specialisation necessitate small-scale organisation of production as well as greater fluidity in collaborative connections. The network character of organisational relations becomes stronger, and the boundaries of and between organisations become less visible or important.

Technological developments cause and facilitate these changes. Production demands becomes more knowledge intensive and consequently more professional; knowledge becomes more and more a core economic commodity by which professionals – as producers of knowledge – are acquiring an increasingly prominent economic position. Knowledge-intensity and professionalisation, in turn, strengthen the trend to small-scale organisation in which the smallest organisational units become more autonomous and, at the same time, more interwoven in horizontal networks (with co-producers, suppliers, contractors etc.).

It is these patterns of relationships which, both empirically and analytically, are acquiring greater importance than the classic, vertically integrated and externally delimited organisations as physical entities.

> At the centre of analytical focus one finds the cultural and institutional frameworks which facilitate the diverse forms of calculation and modes of rationalities within which are constituted networks of organizational relations. The actual form that organizations take occurs within the specification of these features. (Clegg, 1990: 152)

The plurality of organisation forms is growing; the large bureaucracy and the large business concern are no longer the dominant form, although of course extensive bureaucracies, conglomerates and multinationals continue to exist and exercise world-wide influence. However, because the small-scale is a technological option – which historically is relatively unique – and because the market demands growing flexibility, the economic dynamic of 'economies of scale' changes to one of 'economies of scope' (Crook *et al.*, 1992: 179). Bureaucratic hierarchies are then less adequate. They disappear as a pattern of organisation; they are replaced by infrastructures and architectures of information provision and communication. The infocracy is the new bureaucracy (Zuurmond, 1994).

Changes in organisation patterns, professionalisation and technological developments are a part of the dynamic of capitalism, and as such of the modernisation process (Crook, *et al.*, 1992: 29–30). On the other hand, however, there are signs of a break with the modernisation process: in the obsolescence of functional, specialised and differentiated bureaucracy, in the pluralism of organisation patterns, and in the autonomisation of professional expertise.

Both in the area of economic transformation, and in that of organisational change, we see the growing significance of the cultural factor: in consumption, the plurality of lifestyles, knowledge and professionalism. Recognising the importance of cultural factors has always been the great contribution of cultural sociology (Weber, 1985 (1922); Zijderveld, 1983). In that interpretation, modernisation is a cultural process of advancing functional rationalisation. The diagnosis of many postmodern or postmodernist writers, however, is quite different: they observe, not cultural generalisation, but pluralisation. Modern culture is fragmenting and is ending up as a fluid variegation, whose significance is its penetration of other spheres and its autonomy. Processes of individualisation, secularisation and depillarisation are as much cause as effect. The crushing of a collective consensus based on a few great ideological narratives, and their replacement by a culture of calculation, strengthen this variegation on the one hand, and on the other, bear witness to continuity of the hegemony of the capitalist ethos.

The break with the modernisation process lies primarily in the phenomenon of mass culture, which has both displaced and absorbed 'high culture' as *avant-garde*, and of plurality of lifestyles, which creates identities. Identities, incidentally, which are the result of choice and seduction, but which are constantly changing and temporary. In this, consistency is no longer required and eclecticism is entirely acceptable. Modernism as the cultural norm, with its distinction between high and low culture and its acceptance of the *avant-garde* as representing progress is the official culture (Foster, 1985), the canonisation of good taste. Postmodernism as a cultural variant, 'ironises' modernism and replaces it with collage, pastiche and deconstruction. Lacking a belief in progress, it has no room for the avant-garde. Tradition is only an archive (Crook, *et al.*, 1992: 65) from which one can draw freely and from which eclectic lifestyles can be chosen. The core of postmodernity as a cultural pattern is, according to Harvey,

> (...) its total acceptance of the ephemerality, fragmentation, discontinuity, and the chaotic (...). It does not try to transcend it, counteract it, or even to define the 'eternal and immutable' elements that might lie within it. Postmodernism swims, even wallows, in the fragmentary and the chaotic currents of change as if that is all there is. (Harvey, 1989:44)

It is precisely this characteristic of postmodernism and postmodernisation which seems to be penetrating other spheres such as the economy, organisations, technology, administration and politics. And even though in many domains we may be able to speak of advanced modernisation, or perhaps reflexive modernisation, as do Beck, Giddens and Lash (1994), precisely because of the pre-eminence of the cultural factor in a society like ours, a pattern of postmodernisation can not be excluded.

In Front of the Decor: Public Administration, Technology, Politics

That conclusion – the possibility of postmodernisation as the pattern of change being experienced by our society – is based, on the one hand, on the description of the decor of social developments in this chapter. On the other hand, however, it can also be traced back to previous chapters. I shall summarise the most important observations and analyses.

In public administration, an ambiguous set of developments is visible. First of all, there are amendments to classic bureaucracy:

- There is a constant process of reorganisation, designed to make the bureaucracy more flexible and adaptable to complexity and turbulence in its environment.
- Adapting and refining the policy instruments are a part of this: instrument theory has a prominent place in the academic discourse of public administration.
- Being business-like is an important value which gives direction to the processes of change. Efficiency and effectiveness are orientations which in a sense compete with the classic rationalities of power and law.
- All these amendments to bureaucracy can be seen as attempts to (further) rationalise the bureaucracy. The discrepancy between ideal-typical and empirical manifestation should be removed. It entails modernising public administration.

In the second place, there is the question of differentiating the bureaucracy. The classic conception of a machine, which serves as a tool for its political masters, is breaking up:

- By a process of localisation, the bureaucracy is brought 'closer' to social processes. This involves forms of decentralisation and regionalisation which take bureaucracy out of the domain of the state and into the domain of local and regional administration.
- Functionalisation of bureaucracy mainly involves making sections of it independent. Political control is reduced and orientated differently. Professionalism and neo-corporatist involvement become important criteria for its design and methods of work.
- In general, we see a pattern of increased autonomy for parts of the bureaucracy and for the professionals and managers within them. Both territorial and functional decentralisation break through the image of monolithic bureaucracy.

- As forms of bureaucratic differentiation, these developments are to be understood as advanced modernisation. Yet at the same time, they undermine the classic conception of coherence and vertical integration which distinguishes modern bureaucracy. In that sense, they limit the significance of modernisation.

That last point, and my third, is even more relevant to a number of reversals in bureaucracy:

Localisation, functionalisation and autonomisation of bureaucracy are replacing the image of the pyramid by that of the archipelago. Naturally, this is an ideal-typical image of bureaucracy, which is empirically always modified. Recent developments, however, are extending the variety and plurality of organisational patterns to such a degree that the adequacy of the pyramidal model – as a guideline for designing the politico-administrative system – has become problematic.

- Processes of devolving independence appear, it is true, to relate primarily to policy implementation, which remains hierarchically subject to policy development and definition. Nevertheless, there is a discernible change. If the theory that policy is actually created in the stubborn complexity of implementation is plausible, independent organisations will increasingly become the real producers of policy. Huigen (1994) speaks of strategic policy implementation.
- Organisations in public administration are becoming more autonomous, and more interwoven into varying patterns of contractual ties with other – public and private – organisations and citizens. That too increases the fragmentation and disintegration of the pyramid.
- These reversals of bureaucracy represent a break with tradition. They reveal a kinship with what others have described as postmodern developments in organisations (Bergquist, 1993; Clegg, 1990; Cooper & Burrell, 1988; Hassard & Parker, 1993; Nooteboom, 1992; Parker, 1992; Thompson, 1993).

In the area of technological development, we can also see diverse patterns: digital ambiguities. Information and communication technology – the dominant technology in public administration and politics – has as its main characteristics:

- growing capacity, both quantitative and qualitative, of systems and applications, which make reductions of scale and greater autonomy technological options;
- expansion of networks which facilitate communication, virtually independently of time and space, and create many horizontal connections;

- the coupling of databases, systems and applications, by which profiles can be constructed for pro-active as well as repressive ends;
- development of virtual reality, by which realities can be constructed and simulated.

These characteristic developments in the field of ICT are of fundamental organisational and cultural importance.

Organisationally, a number of consequences have already been distinguished:

- ICT is a technology which contributes to the arsenal of controls in modern organisations. It embodies the myth of perfectibility: thanks to this intellectual technology, we know more and more, and can create an almost perfect set of instruments for organisations.
- Through ICT, organisations are becoming continually more transparent: the machine is being refined and turned into an ever more flexible instrument for the exercise of power. At the same time, the environment is also becoming more transparent: citizens, clients and markets. The electronic traces which each leaves behind, and the electronic databases at the disposal of organisations, are making increasingly intricate interventions and operations possible.
- Through ICT, horizontal relationship patterns within and between organisations are acquiring far greater importance that the classic vertical relationship patterns of bureaucracy and hierarchy. The explosive success of the Internet demonstrates this development almost ironically: military in origin, anarchistic in conception and operation, extremely effective in terms of the capitalist market.
- Through the symbiotic connections between people, organisations and ICT, a virtualising effect has been created. Computer generated realities are replacing other realities and are acquiring unprecedented meaning and significance. The attraction of ICT is great; the hype realistic.

In a cultural sense, ICT is ambiguous.

On the one hand, ICT is an unambiguous artefact of modernisation. The dominant code of meaning attached to, and even embodied in, ICT is functional rationality. Extending ICT in organisations is a form of rationalisation, bureaucratisation and differentiation. The potential for control is expanded, and social and psychical reality are added to the objects of control.

On the other hand, developments in ICT have ushered in such transformations that, in the eyes of many writers, we should speak of a process of postmodernisation. For one thing, there are the changes induced by ICT in time and space, and our experience of them. The progressive compression of time and space remove them as obstacles (see Harvey, 1989: 284ff.): the world is a village, a village the world. There we see simultaneously cultural globalisation and cultural fragmentation. For another, ICT reduces subjectivity. Man, as the deciding subject with the most advanced powers, disappears from the centre: technology is no longer a tool, but a relatively autonomous culture. The subject is thereby de-centred. His identity is more and more dependent on technologies and is increasingly determined by his fragmented presence in systems, files and applications (see also Poster, 1990).

A variety of connections exist between developments in public administration and technology. These connections are often contingent: in practical or academic discourse, change, experiment and innovation in public administration are relatively seldom tied to ICT. They are usually part of separate strategies of change and modernisation. Both outcomes, however, become unmistakably involved with each other in the actual design and working methods of the administration.

In the first place, there is the classic association between informatisation and bureaucratic culture. The characteristics of informatisation in the mainframe era were standardisation, formalisation, specialisation and centralisation. Here we have a Weberian 'Wahlverwandtschaft', an affinity between technology and bureaucratic culture. ICT contributes to the process of modernisation and bureaucratisation (see Frissen, 1989).

That contribution, however, is specific, because informatisation in bureaucracy strengthens, in particular, its technocratic potential. Expertise, information and knowledge become more prominent as sources of power, in respect of the traditional political sources of power.

Zuurmond (1994) goes a step further in arguing that the special combination of ICT and bureaucracy, produces a new ideal-typical configuration: that of infocracy. There classic bureaucracy disappears in the virtual reality of infrastructures and architectures of information provision and communication. The organisation even 'de-bureaucratises'; it becomes more flexible and parts of it become more autonomous, although still within the rules, norms and standards incorporated within the infrastructure.

But it is still possible to speak of modernisation. Even if informatisation is introduced in the hope of change, experiment and innovation. ICT is able to improve considerably the client-orientation of the bureaucracy. But behind the friendly face of Soft Sister there lurks inevitably the ugly mug of Big Brother: tailor-made services require tailor-made controls and surveillance. ICT could even lead back to greater steering ambitions, especially since the availability of so much information about policy implementation can soon lead to anticipatory policies.

But at the same time, opposite developments are afoot. The combination of network technologies and independence leads to other relationships between traditional policy departments and their executive organisations. Policy-making then could well become the result of processes of information exchange and communication between independent units and their particular environments. Policy networks have, after all, a digital form. The same network technologies can be introduced into processes of (direct) democratisation, through which horizontal patterns of deliberation and decision-making can emerge.

Administrative and technological developments together lead to transformations which contain the possibility of postmodernisation:

- Horizontalisation: both administrative and technological relationships acquire a network character, which modifies the classic pyramid.
- Autonomisation: administrative units become more autonomous, while technological developments and characteristics in turn facilitate that autonomy.
- De-territorialisation: in particular, technological development makes the importance of territory as the rationale for structuring an organisation arbitrary. That leads to an administrative revolution, because the politico-administrative system is still based on territory.
- Virtualisation: ICT-generated meaning in public administration is constantly gaining in influence, by which 'the' reality and its definitions are acquiring a more virtual character.

All these developments are particularly problematic for the position and significance of politics. They all attack the classic notion of political primacy. That is already an immanent tendency within the process of modernisation, in which bureaucracy, and especially technocracy, attempt to push politics aside. This tendency becomes radicalised in processes which indicate postmodern transformations. In these, decision-making processes become flatter, the pyramid collapses into an archipelago, and the territory becomes irrelevant. Classic institutional politics becomes problematic because it ends up in a vacuum. From this also arises the futile romanticism of communitarianism and republicanism. As do the pleas for a restoration of the primacy of politics. And the disappointing attempts at institutional reflection.

Can the empty place of power be identified? That question will be addressed in later chapters. But before doing that, it will be necessary to consider further some of the themes in postmodernisation. We shall do that in a 'theoretical intermezzo'.

7 Theoretical Intermezzo

Postmodernism is a fashion, it is not new, it is irresponsibly relativist, it is anti-rational, it is eclectic, it is a 'yuppy-philosophy', it is vulgar, it is superficial, it is anti-humanist, it is reactionary. From the point of view of its critics, all these descriptions are undoubtedly correct.

Postmodernism is fashionable: fashion is, after all, seen as an important phenomenon in a fragmented culture of innumerable lifestyles. Postmodernism is not new: the modernist ambition to be *avant-garde* is a futile one and can happily be abandoned. Postmodernism is irresponsibly relativist: 'disbelief in meta-stories' (Lyotard, 1987a: 26) is an important starting-point, which since it is not a new narrative does not need justification. Postmodernism is anti-rational or irrational: the singular rationality of the 'man-made order' has finally led to a 'long march to prison' (Bauman, 1992: XVII); the irrational is anything which strategies of discipline have defined as beyond normality. Postmodernism is eclectic: culture and tradition constitute an archive from which one can draw at will to satisfy a hedonistic aesthetic. Postmodernism is a yuppy-philosophy: consumption patterns create cultural and social identity, and not the reverse – a certain level of prosperity has made postmodern chaos acceptable. Postmodernism is vulgar: there are no longer any cultural taboos – Marco Borsato is just as acceptable as Karl Heinz Stockhausen; Disneyland as acceptable as Alhambra; and Rob Scholte creates a work of art out of a car wreck in which he was blown up. Postmodernism is superficial: 'if one's moral identity consists in being a citizen of a liberal polity, then to encourage light-mindedness may serve one's moral purposes. Moral commitment, after all, does not require taking seriously all the matters that are, for moral reasons, taken seriously by one's fellow citizens. It may require trying to josh them out of the habit of taking those topics so seriously' (Rorty, 1991a: 193). Postmodernism is anti-humanist: humanism is narcissistic and anthropocentric – man is no longer the centre of the world and the measure of all things; man is de-centred (Poster, 1990: 7). Postmodernism is reactionary: progress or improvement in terms of an overarching ideal – a grand narrative – is not possible; there only remains the possibility of muddling through within the status quo.

However: a single 'postmodernism' does not exist. That would not be possible, since it would resurrect the metaphysics of the grand narrative. So a friendly smile is probably the best reaction to all those critical reactions. After all, anything goes. Rorty again:

> If we take care of freedom, truth can take of itself. If we are ironic enough about our final vocabularies, and curious enough about everyone else's, we do not have to worry whether we are in direct contact with moral reality, or whether we are blinded by ideology, or whether we are being weakly 'relativistic'. (Rorty, 1989: 176)

Being ironic, however, is difficult enough, particularly if one wants to write a 'serious' book. And a systematic survey of postmodernist insights is utterly impossible. Being systematic is fatal for irony. Nevertheless, I shall risk an attempt to give some idea of postmodern vocabulary. Although there is great variety, it is possible to distil a number of ideas out of postmodern literature and the literature of postmodernism. It will not be an exhaustive survey: I shall rely on a few writers and choose those ideas which I can use in the context of this book's themes. Philosophically, I have been most inspired by Rorty because of his light touch, irony and humour – rare qualities in a world populated by former Marxists and gloomy Frenchmen. For the social sciences, I rely particularly on the work of Crook, Pakulski and Waters (1992) together with Bauman (1992), Harvey (1989), Lash (1990) and Poster (1990).

7.1 POSTMODERNISATION, POSTMODERNITY, POSTMODERNISM

'All Cretans are liars', said the Cretan. 'Truth does not exist'. Classic philosophical problems are distinguishing features of postmodern literature. For instance, should a scientific analysis of postmodernism, itself have a postmodern orientation? Bauman thinks it should:

> Postmodernity (or whatever other name will be eventually chosen to take hold of the phenomenon it denotes) is an aspect of a fully fledged, viable social system which has come to replace the 'classical' modern, capitalist society and thus needs to be theorized according to its own logic. (Bauman, 1992: 52.)

Crook *et al.* are critical of this position. In the first place, it is premature to assume a transition from modernity to postmodernity. In the second place, it is doubtful if postmodernity constitutes a coherent social system in the way that modernity does (Crook *et al.*, 1992: 232). But the converse of that view is also problematic. The view, encountered in Clegg (1990), Featherstone (1988) and Lash (1990), that a sociology of the postmodern will suffice, misunderstands the deep connection between sociology and modernity. Putting it more pointedly, is not every analysis of the postmodern, which attempts to provide a consistent and coherent narrative, an internal contradiction and modernist in nature?

Or should we follow Rorty, who resolves the dilemma by distinguishing clearly between the private and the public sphere? Being postmodernist is a private event. Irony – as characteristic of choice – is an unsuitable attitude for the public sphere. He suggests:

> that irony is of little public use, and that ironist theory is, if not exactly a contradiction in terms, at least so different from metaphysical theory as to be incapable of being judged in the same terms. (...) Ironists should reconcile themselves to a private-public split within their final vocabularies, to the fact that resolution of doubts about one's final vocabulary has nothing in particular to do with attempts to save other people from pain and humiliation. (...) We should stop trying to combine self-creation and politics, especially if we are liberals. The part of a liberal ironist's final vocabulary which has to do with public action is never going to get subsumed under, or subsume, the rest of her final vocabulary. (Rorty, 1989: 120)

Elegant as this solution may be, I am not happy with it. The possibility of an ironic theory of politics and administration is too alluring. I shall attempt to formulate such a theory – though not as any grand narrative – later in this book.

Also, for the moment, I shall happily leave unanswered the question whether an analysis of the postmodern should itself be postmodern. It is a question of self-preservation and conviction: I have no other instruments of analysis and I am only able, and only willing, to tell a single narrative. In such cases, fantasy is often more important than methodological validity. Whoever wishes to be charmed or convinced are welcome; those who do not can switch to another narrative.

In spite of these qualifications, it is sensible to clarify some of the concepts involved. By that I mean my use of specific terms, rather than an exact definition of them; the terms in question being postmodernisation, postmodernity and postmodernism. Writers such as Featherstone (1988: 197), also make a distinction between them.

Postmodernisation is a complementary concept to modernisation. It indicates specific social processes. According to Crook *et al.*, postmodernisation is a radical extension, or effect, of two modernising tendencies: on the one hand, ongoing differentiation and specialisation which generate such a high level of complexity that problems of integration arise; on the other, rationalisation and commodification which have led to such a high level of organisation that an internal contradiction arises with the process of differentiation (Crook *et al.*, 1992: 32–35). Postmodernisation is therefore a process which displays continuity with modernisation: it is radicalisation of tendencies, already present, such as unpredictability, chaos, hyper-differentiation and autonomisation. Many writers see the cultural factor as increasingly prominent in the process of postmodernisation.

> On a more abstract level the progressive differentiation of culture, society, and personality characteristic of modernity involutes so that the very idea of an independent, purely social structural realm no longer makes sense. Rather 'society' must be understood in terms of 'culture' as patterns of signs and symbols penetrate and erode structural boundaries. (Crook *et al.*, 1992: 35)

Or as Jameson puts it with some disdain:

> (…) a prodigious extension of culture throughout the social realm, to the point of which everything in our social life – from economic value and state power to practices and to the very structure of psyche itself – can be said to have become 'cultural' in some original and as yet untheorized sense. (Jameson, 1984b: 87)

While postmodernisation refers primarily to processes, *postmodernity* refers more to a situation. It refers to a historical period which qualifies as the successor to the period of modernity.

> (T)o speak of postmodernity is to suggest an epochal shift or break from modernity involving the emergence of a new social totality with its own organizing principles. (Featherstone, 1988: 198)

Whether the postmodern period has arrived has, according to Featherstone, to be shown by empirical research and analysis. The intellectual, and often merely fashionable, experience of the postmodern is an inadequate barometer. It is often expressed in the French term 'postmodernité', comparable to 'modernité', which reflects rather a specific feeling which a particular class or group has about life, than any historical and empirical phenomenon (Featherstone, 1988: 200).

Postmodernity as a period may be characterised as post-industrial, as post-Fordist, with a specific cultural logic in which image predominates, where the primary political unit is no longer the national state and ICT is the dominant force of production. At the same time, it is clear that observing postmodernity is not a neutral activity. The observer always takes an implicit position. One has, therefore, always to remain sensitive to that position and the relational character of every sociological analysis of postmodernity. Without being modern, the transition from modern to postmodernity can not be sketched. But we must also be postmodern if we are to avoid describing postmodernity as a variant of modernity, and thus remaining a part of the meta-narrative of modernism.

That last position is theoretical or philosophical in nature, but also has links with the world of art, architecture, literature, cinema etc. It is a position of *postmodernism*. Harvey describes it well:

> (...) postmodernism, with its emphasis upon the ephemerality of *jouissance*, its insistence upon the inpenetrability of the other, its concentration on the text rather than the work, its penchant for deconstruction bordering on nihilism, its preference for aesthetics over ethics (...). (Harvey, 1989:116)

It is a philosophical position which rejects every form of metaphysics and thus connects such widely different writers as Nietzsche, Heidegger, Derrida, Lyotard, Baudrillard and Rorty. It is a relativist position which denies that there is a higher truth, that human beings are fundamentally moral, or that history has a goal or embodies progress. The fragmented culture of postmodernity has its counterpart in the theoretical or artistic positions of postmodernism. In it we find a preference for eclecticism, for pastiche and cliché, for images and signs, for mass culture, and for their reflection in non-normative, non-totalising, and non-foundational (short) narratives. Postmodernism is superficial with conviction (Rorty, 1991a: 181) Simulation is no longer a reflection or representation of an 'actual' reality, but reality itself (Baudrillard, 1985: 127).

Having defined the concepts, I shall develop them further in the rest of this chapter.

7.2 WE POSTMODERN BOURGEOIS LIBERALS

Richard Rorty, from whose work (1991a: 199) I have borrowed the title of this section, is an exceptionally attractive philosopher. He is elegant, humorous, accessible and surprising. He lacks the continental melancholy of the French postmodernists Derrida and Lyotard and is less exuberant in his choice of words and use of language than Baudrillard. Nor does he have to wrestle with a Marxist past, unlike so many English and American writers for whom postmodernism – surprising as it may seem – is a new political ideology. He is also a stranger to that outgrowth of vulgar postmodernism, 'political correctness'.

For these reasons, I have chosen to follow Rorty in my treatment of fragments of postmodern thinking. We are not of course talking about foundations - they have been rejected - but ways of reasoning and of establishing one's position. Postmodernist convictions, as Nietsche said, are no worse enemies of truth than lies. Postmodernist convictions are contingent, just as is truth, and postmodernism continually emphasises that contingency.

The Grand Narratives

'With extreme over-simplification we can describe the 'postmodern' as disbelief in meta-narratives', as Lyotard concisely puts it (1987a: 26). Modernity is the era of the great ideologies of Liberalism, Communism, Social Democracy, and Christian Democracy. All these ideologies present a grand narrative, a narrative which is consistent, which legitimises behaviour, which explains and explains away contingencies, and which understands the direction of history. The Christian Democratic narrative aims to combine the premodern with the modern. The other narratives are entirely modern: they see themselves as a part of progress and ultimately as a contribution to human emancipation. They are rational and scientific: rationality and science constitute the legitimising core of their ideology.

All grand narratives claim to understand modernity and regard themselves as an expression and actualisation of it. Of course, modernity is an uncompleted project (Habermas, 1985b), but its completion is both inevitable, since that is buried within the process of modernisation, and a moral imperative, since it is inextricably linked to reason which, while empirically specific, can boast a universal validity (Habermas, 1987).

The fiasco of the grand narrative is exposed, for many postmodernists, in two major historical events: the Second World War and the genocide of the Jews. 'Auschwitz' can be seen as a paradigmatic name for the tragic 'incompleteness' of modernity' (Lyotard, 1987b: 27). And then there is the failure of communism as a fundamental alternative and geo-political threat to capitalism and democracy. From now on, ideological struggle appears to be confined to variations on the theme of liberalism and capitalism.

But it is not only the ruthless narratives of fascism and communism which have fallen short. Their totalitarian character is evidence which few will contest. But the core of the postmodernist position is that all ideologies are totalitarian. They all claim to know the 'true' nature of social relationships, how to resolve conflict, and how to achieve social progress. In that sense, they are 'totalising' and 'terroristic' because they legitimise the destruction or removal of differences (Larrain, 1994: 299). This applies even more because many ideologies present themselves as critical of ideology, as true narratives able to unmask the ideological pretence of competing narratives.

In the words of Baudrillard:

> It is always the aim of ideological analysis to restore the objective process; it is always a false problem to want to restore the truth beneath the simulacra. This is

ultimately why power is so in accord with ideological discourses and discourses on ideology, for these are all discourses of truth – always good (...) (Baudrillard, 1988: 182).

In other words, there is no objective criterion by which to judge the truth of ideologies. They are meta-narratives which exist next to and in opposition to each other. Nor is there any normative criterion by which to judge the moral content of ideologies. Rorty stresses time and again that such a criterion is unattainable, undesirable and unnecessary. We can be satisfied with the institutions of liberal democracy, without having to provide them with an ideological foundation. In fact, it is better *not* to do it.

> The advantage of postmodernist liberalism is that it recognizes that in recommending that ideal [the idea of liberal, procedural justice, PF] one is not recommending a philosophical outlook, a conception of human nature or of the meaning of human life, to representatives of other cultures. All we should do is point out the practical advantages of liberal institutions in allowing individuals and cultures to get along together without intruding on each other's privacy, without meddling with each other's conceptions of the good (Rorty, 1991a: 209).

Nietzsche and Heidegger Completed

The end of the grand narratives is not just proclaimed by postmodernists as an empirical 'fact'. Bell did that in *The End of Ideology* (1960) and did so with some regret because it meant the disappearance of 'bourgeois values of classical capitalism' (Turner, 1989: 203) and that the system would become entangled in its cultural contradictions (Bell, 1976). Postmodernists advocate the end of the grand narrative, because they want to be anti-metaphysical. They are more or less completing the philosophical mission of Nietzsche and Heidegger, both of whom wanted a philosophy without a metaphysical or moral foundation.

This continuity is apparent in the work of Foucault, Derrida, Lyotard and Baudrillard, all of whom are inspired by Nietzsche and Heidegger. It also applies to Rorty who adds the empiricism of Dewey and Davidson as important sources. Postmodernism is anti-metaphysical because it does not believe in the possibility of final grounds, essences, deeper criteria or foundations. There are only vocabularies or language games, which have meaning for individuals and groups through traditions and common experiences. And although every individual or group can have a specific 'final vocabulary' which legitimates utterances and behaviour, there exists no meta-vocabulary that enables us to assess every other vocabulary either objectively or normatively (Rorty, 1989). Language, after all, is contingent and therefore so is the subject. Philosophy is a debate about possible descriptions of the world, conducted in a particular manner.

> But 'argument' is not the right word. For on my account of intellectual progress as the literalization of selected metaphors, rebutting objections to one's redescriptions of some things will be largely a matter of redescribing other things, trying to outflank the objections by enlarging the scope of one's favorite metaphors. So my strategy will be to try to make the vocabulary in which these objections are phrased look bad, thereby changing the subject, rather than granting the objector his choice of weapons and terrain by meeting his criticisms head-on. (Rorty, 1989: 44)

The anti-metaphysics of postmodernism rests partly on an epistemological anti-representationalism. There is no empirical reality, independent of our vocabularies, which is represented in those vocabularies. Descriptions and redescriptions of the world do not exist next to or in opposition to reality; they are the reality. Meanings are central; objectivity can not be demonstrated (Gellner, 1992: 24, 35). The actual thereby becomes problematic because it only exists in the description of it. Representation overshadows reality (Lash, 1990: 15). Baudrillard speaks of 'hyperrealism' to describe the simulation that has become reality (Baudrillard, 1985: 128) or 'The map that engenders the territory' (Baudrillard, 1988: 166). Nor, in postmodern eyes, does the totalising rationality of modern science deserve any mercy (Jackson & Carter, 1992: 12), a standpoint which – curiously enough – finds support from the reformationist philosopher Geertsema (1988). For that reason, intellectuals are no longer 'legislators', but 'interpreters' (Bauman, 1992: I ff.).

Anti-metaphysics and anti-representationalism are themselves targets of criticism. Habermas fulminates regularly against the rejection of reason and progress as humanist idealism, which result from the postmodernist credo. Marxists reject anti-utopianism and the criticism that the struggle for emancipation is just a 'dream of humanity'. Criticism of the Enlightenment's pretension to rationality and objectivity must itself be rational and objective if it is to be convincing, argues Callinicos (1989: 80ff.). And Harvey accuses postmodernism of making coherent politics impossible and opening the door to reactionary neo-conservatism (Harvey, 1989: 116). According to Gellner (1992), one of its fiercest critics, postmodernism is intolerably and unacceptably relativist. But in that, postmodernism is reproached for something which it opposes:

> Relativism certainly is self-refuting, but there is a difference between saying that every community is as good as every other and saying that we have to work out from the networks we are, from the communities with which we presently identify (...) The view that every tradition is as rational or as moral as every other could be held only by a god, someone who had no need to use (but only to mention) the

terms 'rational' or 'moral', because he had no need to inquire or deliberate. Such a being would have escaped from history and conversation into contemplation and metanarrative. To accuse postmodernism of relativism is to try to put a metanarrative in the postmodernist's mouth. One will do this if one identifies 'holding a philosophical position' with having a metanarrative available. If we insist on such a definition of 'philosophy', then postmodernism is postphilosophical. But it would be better to change the definition. (Rorty, 1991a: 202).

Moral judgements are part of every vocabulary. But we can not choose between vocabularies on the ground of some superior criterion, nor on the criterion that they are all equally valid.

Aesthetics Instead of Morality

Morality in postmodernism appears to be giving way to aesthetics. Sociologically, this can be explained by the growing prominence of the cultural factor in postmodernity. Images, styles, consumption patterns, form cultural identities and provide the norms of a 'good' which is, however, fragmented. The good is thus, more often than not, self-selected beauty or aspiring to what is pleasurable. Bell speaks of hedonism and narcissism as characteristics of the postmodern. Burrell says that in postmodernism, pleasure is more important than seriousness and that desire should be taken more seriously than truth (Burrell, 1993).

Postmodernism rejects the idea of foundations or essence. There is no 'Archimedean point' which can count as the non-historical essence (metaphysically or biologically) of morality and ethics (Shusterman, 1988: 339). So the good life can only be formulated within a contingent vocabulary. But vocabularies are historical and pluralistic. The good life, then, is not a normative phenomenon to be discovered, but a historical construction to be made and formed (Shusterman, 1988: 341). What remains is aesthetics, taste. Moral judgements are a part of a larger conglomeration of possible judgements which define the just and the good. The choice of judgement is not essentially moral, but a historical contingency, the result of the struggle for self-realisation and identity-creation. In that process, according to Rorty, irony is an important guideline.

> The liberal ironist just wants our *chances of being kind,* of avoiding the humiliation of others, to be expanded by redescription. She thinks that recognition of a common susceptibility to humiliation is the *only* social bond that is needed. (Rorty, 1989: 91)

Rorty advocates 'light-mindedness' in making moral judgements. Anything which goes beyond that is potentially dangerous, since it will be totalitarian and essentialist.

> I can, however, make one point to offset the air of light-minded aestheticism I am adopting toward philosophical questions. This is that there is a moral purpose behind this light-mindedness. The encouragement of light-mindedness about traditional philosophical topics serves the same purposes as does the encouragement of light-mindedness about traditional theological topics. Like the rise of market economies, the increase in literacy, the proliferation of artistic genres, and the insouciant pluralism of contemporary culture, such philosophical superficiality and light-mindedness helps along the disenchantment of the world. It helps make the world's inhabitants more pragmatic, more tolerant, more liberal, more receptive to the appeal of instrumental rationality. (Rorty, 1991a: 193)

Disbelief in grand narratives, resistance to metaphysics and representationalism, objections to a non-aesthetic morality, are the political positions of postmodernism. Rorty certainly defends a liberal society as the most attractive kind of society. But in that defence, seduction is a more important instrument than argument or any normative foundation. For any attempt to provide a normative foundation should be mistrusted. Postmodernism does not supply foundations; it only tells many short stories.

7.3 YUPPIES, EPHEMERA AND KITSCH

If philosophical postmodernism is part of the intellectual baggage of bourgeois liberals, as Rorty ironically and well-meaningly suggests, then we can suggest, equally ironically and well-meaningly, that postmodernity is the era of yuppy-culture. The preference for ephemera over stability, fragmentation over coherence, pastiche over a strict style, is elitist. Consumerism as the primary determinant of identity often demands more than merely fashionable purchasing power. A large city is only an attractive decor for the owner of a penthouse, not for someone living in rundown accommodation in an impoverished district. Designer drugs and house parties may be postmodern, but what about drugs like crack?

The answer ought to be that they are all postmodern, if we are consistent in our assertion that postmodernism is not a grand narrative, takes no moral stance, and can not, in any way, become a new ideology of liberation or 'political correctness'. Refusing to acknowledge foundations or essences because they are dangerous, even morally, implies a refusal to predict a better society, a more attractive culture, and less cruelty. It remains, of course, possible to wish for a better society, higher culture and less cruelty, but that wish is contingent and, in social terms, pluralistic.

So what is the cultural significance of postmodernism, and what could be the cultural characteristics of postmodernity? What cultural changes are implicit in the process of postmodernisation?

Elites

Cultural change and renewal always involve elites. For cultural modernism, being elitist is an important goal because it wants to be *avant-garde*. And however oppositional cultural modernism may seem, for the moment it is the official culture, conserved in museums, taught in universities, crystallised in design and architecture (Foster, 1985: IX). Crook *et al.* (1992) refer to the 'syndrome of cultural modernity'. This syndrome is the result of a variety of processes, which have reached a certain level of intensity and contain within them the seeds of postmodernisation. They are: differentiation, rationalisation and commodification. In the first place, there is the differentiation of the cultural, in respect of other domains. Such differentiation reaches a point where it has advanced so far – hyper-differentiation – that any differences from other domains becomes irrelevant. In the second place, rationalisation involves the autonomous development of cultural and other values, as well as the development of production and reproduction techniques. Rationalisation reaches its limits as modernisation becomes more of a 'museum culture', as the *avant-gardes* cease to be *avant-garde,* and tradition slowly erodes. Commodification undermines the modernist distinction between 'high' and 'mass' culture because there is no cultural domain which is immune from commodification. 'Taste' is no longer the preserve of dominant social groups (Crook *et al.*, 1992: 56–57).

Culture as a separate domain does not disappear, but it does penetrate all the other social domains. ICT gives production and reproduction a greater impetus. The combination of ICT and classic mass media produces an entertainment culture (hence 'infotainment') on a global scale which, unlike modernist culture, is non-hierarchical (Bauman, 1992: 31). The proliferation of lifestyles, typical of postmodernisation and based on consumption patterns, gives rise to a pluralisation of culture which is no longer tied to class (Heller & Fehér, 1988: 133ff.).

According to Lash, a mark of postmodern culture is a reduction in the autonomy of the cultural domain because it penetrates other domains and thereby becomes indistinguishable from the social; because the boundaries between cultural production, consumption and distribution become diffuse; and because the relationship between representation and reality as 'mode of signification' becomes problematic. Modernism problematicises representation; postmodernism problematicises reality (Lash, 1990: 11–13).

But although postmodernism in a cultural sense (e.g. architecture; see Jencks, 1986) displays characteristics of the *avant-garde* – provocative, revealing, innovative – , it does not intend to do so. After all, the possibility of an *avant-garde* implies some form of progress, improvement and depth which it rejects as impossible and ideological. And despite the fact that the postmodern design of Alessi is extremely expensive and, for that reason alone, could never become mass culture, postmodernism distinguishes itself by its denial of a 'high-mass' hierarchy within culture. A distinction between kitsch and art can never be founded; there are merely different tastes, different consumption-based identities, different kinds of beauty and ugliness. That means that while postmodernism is non-elitist and non-hierarchical, it is also non-egalitarian. Differentiation and inequality are encouraged, not to introduce order but to reject it. As they say: 'anything goes'.

Pastiche as Choice

Harvey characterises the cultural ethos of postmodernism as follows:

> I begin with what appears to be the most startling fact about postmodernism: its total acceptance of the ephemerality, fragmentation, discontinuity, and the chaotic (...) But postmodernism responds to the fact of that in a very particular way. It does not try to transcend it, counteract it, or even to define the 'eternal and immutable' elements that might be within it. Postmodernism swims, even wallows, in the fragmentary and the chaotic currents of change as if that is all there is (Harvey, 1989: 44).

In architecture (see Jencks, 1986), there is a strong eclecticism and no attempt is made to make a single statement of style. Postmodern buildings are simultaneously classical and modern, while ornament and ornamentation (Lash, 1990: 37) are given prominence. The modernist credo that form follows and represents function is somewhat mischievously rejected. In literature, text and the intertextuality of texts are central. It is arrogant to presume that we control the meaning of the text: 'Language works through us' (Harvey, 1989: 50–51). The power of the producer declines; the market becomes the regulator. Incoherent meanings and sensitivity to market manipulation are the outcome (Harvey, 1989: 51). Postmodernism is a culture of the superficial and the ephemeral. Once the link between representation and reality has been broken, it is only the representations which count. Baudrillard's simulacra are no longer simulations of 'real' reality; they are reality itself – at least in so far as the concept of reality retains any meaning.

Traditions are losing their classic/modern significance. They are becoming an archive.

> The archive can be drawn on in an eclectic and often parodic bricolage of elements (brightly coloured 'toytown' pillars and pediments), or in a 'nostalgic' re-creation of a valued past (Prince of Wales populist and neo-classicism). In either case, the depth and rationality of the modern idea of tradition is lost. (Crook *et al.*, 1992: 65)

Pastiche is therefore no longer some tasteless imitation, but the cultural equivalent of Rorty's irony in philosophy. Originality is no longer possible; everything is a quotation. Identity is found in the specific choice of quotations.

It is not for nothing that modern technology – ICT – is perceived as one of the most powerful forces of postmodernity. The plurality of cultures becomes accessible to all. Knowledge is a commodity which is distributed, moulded and deformed by the new media. McLuhan's statement 'the medium is the message' has only been fully vindicated in the age of the ICT revolution. It is probably the forms of electronic communication, rather than its cultural content, which have a more fundamental cultural impact. Television and the personal computer no longer reflect reality; reality has become a reflection of TV and the PC (Crook *et al.*, 1992: 68). In the previously cited phrase of Baudrillard (1988: 166), it is 'the map that engenders the territory'. I shall return to this in the next chapter.

But the cultural globalisation engendered by ICT and the mass media does not lead to homogeneity but to heterogeneity: local and specific cultures are spread world-wide and generate innumerable specific connections. Tradition becomes an archive, a database, in which all kinds of links lead on to new cultural fragments. Look at fashion, listen to pop music, walk round the great cities: fragmentation is the dominant image. The process of cultural postmodernisation, according to Crook *et al.*, has the following two meanings:

> First, the emerging postculture is not a structure, or a system, or even a syndrome, in quite the way of cultural modernity. It lacks clearly demarcated regions, its boundaries with economy, polity and society are blurred, its hierarchies are multiple and constantly shifting, and it registers no 'depth', no distinction between surface and reality. Second, and in consequence, the effectivity of postculture is of a very different order to that of modern culture. (…) The effectivity of postculture is that of the highly charged particle of meaning, free to move anywhere in sociocultural space and to enter promiscuously into relations with almost any other fragment. It no longer makes sense to ask in general, structural, terms whether culture is a moulder or a mirror of social processes because 'culture' has so pervaded 'society' that the distinction between the two is becoming obsolete. (Crook *et al.*, 1992:75)

Mass Culture and Lifestyles

The prominence of the cultural factor in postmodernity receives perhaps its fullest expression in the phenomenon of mass culture and the proliferation of lifestyles. Cultural postmodernism rejects the distinction between 'high' and 'mass' culture. It is the cultural parallel to the philosophical opposition to the grand narrative, to the meta-narrative. In postmodernity as a historical period, the elitist position of 'high culture' and its associated *avant-gardes* is eradicated by the commodification of all cultural expression and by the massive expansion of ICT and the mass media.

Mass culture is far from homogeneous. Because it embraces all cultural expressions and operates on a global scale, its variety is infinite. This infinity lies in the permanent and contingent plurality of combinations and fragments which mass culture generates. In this, the market is the most important regulator. Through cultural commodification the consumption of cultural products moves more and more to the forefront. Consumerism in post-industrial society destroys the esoteric and elitist elements of classic culture. This consumerism confers identity: a specific choice of cultural fragments, acquired in the market, and combined and diversified into innumerable lifestyles.

If everything in postmodernity is style or form (no depth, only surface; no reality, only appearance), then individuality, or life, also becomes a question of style. One can and one must choose between a completely contingent variety of possible identities. The prescribed frameworks of tradition, religion and ideology have lost their coercive powers, but have not disappeared. They, together with all their associated subcultures, constitute a reservoir of possible practices, identities and repertoires. The grand narrative of national culture is replaced by innumerable short stories of lifestyles. Lifestyles which often seem to be carefully composed, which are compellingly prescriptive and, in their most extreme form as cosmetic surgery, announce the victory of form over content. A victory which appears to make the subject the all-powerful designer of his subjectivity, while, at the same time, the transience and variety of lifestyles point to a decentring of the subject. After all, identity is form, or rather a multiplicity of forms, in which there is no historical continuity or tradition, and which are all borrowed, cited or plagiarised. Lifestyle too is pastiche.

The foregoing is splendidly summarised by Hassan (1985: 123–124) in the following table:

Table 7.1: Modernism and Postmodernism

Modernism	Postmodernism
romanticism/Symbolism	paraphysics/Dadaism
form (conjunctive, closed	antiform (disjunctive, open)
purpose	play
design	chance
hierarchy	anarchy
mastery/logos	exhaustion/silence
art object/finished work	process/performance/happening
distance	participation
creation/totalization/synthesis	decreation/deconstruction/antithesis
presence	absence
centring	dispersal
genre/boundary	text/intertext
semantics	rhetoric
paradigm	syntagm
hypotaxis	parataxis
metaphor	metonymy
selection	combination
root/depth	rhizome/surface
interpretation/reading	against interpretation/misreading
signified	signifier
lisible (readerly)	scriptible (writerly)
narrative/grande histoire	anti-narrative/petite histoire
master code	idiolect
symptom	desire
type	mutant
genital/phallic	polymorphous/androgynous
paranoia	schizophrenia
origin/cause	difference-difference/trace
God the Father	The Holy Ghost
metaphysics	irony
determinacy	indeterminacy
transcendence	immanence

7.4 HYPERDIFFERENTIATION AND FRAGMENTATION

The modernising process is always characterised in sociology as an intensification of specialisation and differentiation. Functional rationality thereby becomes the most important code of meaning. The Weberian 'disenchantment of the world' is a consequence of the gradual victory of functional over substantive rationality. The domain of values, of normative prescription, is overrun by the domain of methods, procedures and means. Tradition and institutions become abstract and their content trivialised. Culture thereby acquires a staccato character (Zijderveld, 1991).

Modernity is a social situation of differentiation and specialisation between and within the various spheres of life: economy, politics, culture, the social and the personal. The functional rationale of systems colonises every social environment, as Habermas argues.

The process of postmodernisation, however, marks a break. The differentiation of modernisation reaches a critical phase of hyperdifferentiation, which leads to all kinds of internal contradictions and culminates in de-differentiation and fragmentation.

Differentiation

Modernisation is a process of differentiation, of '*Ausdifferenzierung*' (Lash, 1990: 172ff.). As well as that, modernisation can be seen as a process of commodification (everything becomes a commodity) and rationalisation (Crook *et al.*, 1992: 7–8). While these processes reveal clear connections, there are also contradictions. For instance, the differentiation of social spheres and rationalisation have led to modifications in the market and the relations of exchange. The irrational consequences of capitalist production relations are compensated for by state intervention. Yet in the end it is the process of commodification which triumphs (Crook *et al.*, 1992: 8).

The most important characteristics of modern society are, on the one hand, complex differentiation between and within domains, and on the other hand, complex organisation. Organisational complexity may be seen as a consequence of continual attempts to control the processes of differentiation through instrumental rationality. Crook *et al.* characterise this society as follows:

Economic features. Large-scale and capital-intensive businesses which produce goods dominate the market by controlling both supply and demand (through advertising). Coherence in the economic system is brought about by financial institutions, and government intervention and regulation. State enterprises are common, while the market as a regulating principle declines in importance. The production of services plays an important part.

Politics. The strongly developed state intervenes widely in social relations through policy and redistribution. Consequently, the state is based on formal democratic principles in which mass parties represent the most important interest groups. In the second place, there is a corporatist consultation structure involving the state, business and unions through which economic and social policy is agreed. Social security is an important source of legitimacy.

Community. The characteristic housing arrangement is the suburb, situated on the edge of an industrial city and typically built to house families. Women stay at home and have caring tasks; men work and are the breadwinners. The inner cities are dominated by offices. These in turn are surrounded by the living quarters of the lower classes. Community formation is often threatened by the plans of businesses and by government interventions.

Cultural patterns. Massification, commodification and rationalisation dominate. The culture industry penetrates the private sphere, particularly through the mass media and electronic communication. They are far from interactive and what they offer is highly uniform. The elite culture has a classical repertoire (music, visual arts, great philosophers, classical intellectuals). Popular culture is considered to be inferior. In this way, culture contributes powerfully to social differentiation.

But within this modern society, important changes are taking place. These changes can be seen as a form of radicalisation in the characteristic process of differentiation and which are referred to by Crook, and others such as Lash (1990), as hyperdifferentiation. In their study 'Postmodernization', Crook *et al.* divide the process of continuous, ongoing differentiation into six areas.

Culture
Modernity has four domains: economy, politics, culture and social community. Culture is subdivided into art, science, morality and law. But this heterogeneity is not the end of the matter. Within each of these 'value spheres' further differentiation occurs as a result of the processes of rationalisation, which are seen as intrinsic, and the commodification of culture, through production and reproduction techniques. The distinction between elite and mass culture tends thereby to disappear (Crook *et al.*, 1992: 16–18).

State
The modern state – corporatist in Europe, 'broker type' in America, totalitarian in communist countries – is a complex and powerful 'intervention machine'. It has a powerful internal orientation towards order and regulation. Its ethos is primarily rational and technocratic. The desire for control is crucial (Crook *et al.*, 1992: 18–20).

Inequalities
In the area of class relations we can observe the following developments: direct ownership becomes more problematic because of the scale of enterprises, by which ownership and control become separated; new middle classes emerge between capital and labour; an important group of professionals arises; the working class is further differentiated; social security produces an underclass. Instead of polarity, there is a strongly differentiated whole of class relations.

Gender-based labour relations change because the state takes over a number of caring tasks, which are subsequently again carried out by women, but now as a paid function (Crook *et al.*, 1992: 20–22).

Politics
Politics has become a separate profession. On the one hand, there is a question of ideology and on the other, there is the daily practice of pragmatism and calculation. The modern politician is impersonal and objective, his ethos increasingly technocratic. The main goal is control by administration which results in extensive bureaucratisation. Legitimacy lies primarily in the mass basis of modern political participation (Crook *et al.*, 1992: 22–23).

Capitalist production
Capitalism is large-scale, monopoloid and largely controlled by managers. The conditions for stability and reproduction – in terms of property, law and social relations – are guaranteed by the state. Technological innovation is of great importance and directed especially at reducing labour costs. Businesses are systematically restructured according to business principles of rationalisation (Crook *et al.*, 1992: 23–25).

Big science
Modern science is also the result of differentiation, rationalisation and commodification. In particular, the combination of science and technology has become a crucial element in production. There are close links between science, commerce and state. Although they constitute autonomous domains, the close links between them create tensions which threaten that autonomy (Crook *et al.*, 1992: 25–26).

Hyperdifferentiation

The advance of differentiation in all these areas results in tensions and contradictions. Indeed, Crook argues that differentiation becomes hyperdifferentiation. The most important contradiction is that between differentiation and the 'monocentric' organisation. Differentiation is no longer a straightforward linear process. It affects both the structure and the function of units in the different domains. Structure ceases necessarily to be a logical consequence of function.

At the same time, modernisation continues to be an expansion of centralised organisation: the vertically integrated pyramid, under unambiguous leadership. But in rational terms, the monocentric organisation is made increasingly dysfunctional by the process of hyperdifferentiation. There are three reasons for this:

- Hyperdifferentiation makes it more difficult to define the sources for control: '(...) there is no longer an oligopolistic set of sources for money and power and they thus lose their effectivity as sources for control.'
- Hyperdifferentiation modifies the potential to control social units because consequences are difficult to predict. 'States become ungovernable, economies unmanageable, and life-worlds anarchistic.'
- Monocentric control assumes a reduction in the distance between the public and the private; it is a form of de-differentiation. But the number of different social objects increases so much that boundaries and distances become blurred (Crook et al., 1992: 34–35).

The result is a process of postmodernisation which, in the various domains discussed above, has the following aspects:

Culture
The ultimate outcome of differentiation is hyperdifferentiation. The internal logic and dynamic of modernity break up, and de-differentiation (Lash, 1990) is the outcome.

> The eventual effect of hyperdifferentiation is to set loose cultural 'fragments' of intense symbolic power which transgress the boundaries between value-spheres and between culture and other subsystems. (...) Postculture offers a flat archive of 'styles' which furnishes materials for pastiche and parody in the place of a developing tradition. (Crook *et al.*, 1992: 36)

The idea of an *avant-garde* disappears, as does the distinction between 'high' culture and 'mass' culture. Cultural patterns, which are now defined by consumer tastes, are no longer hierarchically structured.

State
The social phenomena of de-differentiation and disorganisation have the political effect of decentralising power and separating political conflict from the economic base. The role of the state is lessened by a horizontal redistribution of power between the state and social organisations; by a vertical redistribu-

tion of power to the local level and independent administrative organs; by privatisation and the farming out of state tasks; and by internationalising state tasks.

Class and sex relations
Inequalities are produced more in the sphere of consumption than in the domains of production and the household. Economic power becomes 'more fluid and negotiated' and less firmly attached to individuals, groups and classes. The one-sided distinction between male and female is becoming considerably more differentiated and may even be abandoned (Crook *et al.*, 1992: 38–39).

Politics
Political processes become disconnected from socio-economic relationships. Political differences acquire a conjunctural and contingent character, and relate either to abstract generalised values or to specific lifestyles. Social movements are varied single-issue movements (Crook *et al.*, 1992: 39–40).

Work and production
The structures of organisations are no longer technologically determined. There is a large-scale emergence of small, flexible production units geared to respond as quickly as possible to the market. Many have called this the transition from 'Fordism' to 'flexible specialisation' (Crook *et al.*, 1992: 40).

Science
The hyperdifferentiation and hyperrationalisation of modern science resist all attempts to impose any centralised or externalised co-ordination. Decentring and internationalism are the more probable platforms for scientific development. Furthermore, disciplinary boundaries are becoming blurred. At the same time, the self-evident legitimacy of science has passed, because the struggle for freedom through control has failed.

It is these trends, indicating a process of postmodernisation, to which cultural and philosophical postmodernism gives expression.

De-differentiation and Fragmentation

The term de-differentiation gives expression to a reversal of modernisation. On the one hand, the process of hyperdifferentiation is a radicalisation and universalisation of modernity. We encounter this notion, for instance, in Giddens (1990). The consequence is what Beck, Giddens and Lash (1994) call 'reflexive modernization': the premises of modernity end up in a tangle of unintended consequences, through which modernity gets to contradict itself. According to Giddens, postmodernisation can also be seen as a process of 'modernity coming to understand itself'(1990: 48). On the other hand, de-differentiation could indicate the emergence of new relationships and connections. Lifestyles are one example, but also some of the ideas in communitarism. Lash in particular (in Beck, Giddens & Lash, 1994) attempts to interpret de-differentiation in this way. The grand narratives may have become impossible, but it is important that the short stories resulting from de-differentiation be rooted in a sense of community; which is to say 'a matter of shared meanings' (Lash in Beck, Giddens & Lash, 1994: 162).

The concept of de-differentiation is therefore somewhat problematic. The term 'fragmentation' may be more appropriate. It reflects more clearly the contingency, the potentially ephemeral and the chaotic which distinguish the process of postmodernisation, the postmodern situation and the claims of postmodernism. Admittedly this is a pragmatic rather than a principled argument. Fragmentation is a more convenient concept than de-differentiation, which is not to say that a case for using the term 'de-differentiation' can not be made. As I said earlier, there is no doubt that new connections and relationships are emerging. But in my opinion, their emergence is either contingent or a result of aesthetic choices. Either way, I can not see them as a part of the classical-modern project of progress, enlightenment, emancipation or any other 'totalising' conception.

Fragmentation is a radicalisation of, and in opposition to, differentiation as the dominant characteristic of the modernisation process. In the case of differentiation, there has to be a more or less linear connection and the assumption of a starting point or centre. That also lies at the heart of the idea of progress, which always has a normative connotation. And because progress also has teleological implications – the good is ultimately achievable – the internal contradictions are obvious. Continuing differentiation undermines the linearity of modernisation and results in widespread contingency. It is precisely that contingency which is expressed in the term fragmentation, because it indicates both the accidental nature of social developments and the emergence of new connections. Those connections are the fragments which derive their meaning, not from any grand narrative but from local practices. Many short stories are being told, without a scenario, without direction and without much internal coherence.

7.5   CENTRE AND LINEARITY

The wider pattern of hyperdifferentiation and fragmentation has specific significance for the world of organising and organisations. Because of its relevance to technology, politics and administration, the main topic of the next chapter, I shall devote the following section to the postmodernisation of organising and organisations.

Disorganisation: the Rhizome

Complex, monocentric organisations are pre-eminently a feature of modernity. They form the functional–rational reification of the struggle to control that is the hallmark of modernisation. In a number of respects, they are a reaction to differentiation because they attempt to tame and domesticate the variety and unpredictability to which differentiation gives rise.

Modern organisations combine an ambition for large-scale control with a need for continual achievement. Efficiency and productivity are brought about by technological innovation, whereby modern knowledge-based technology offers previously unimagined opportunities for large-scale control (Cooper & Burrell, 1988: 96). This results in 'the cybernetic-like monolithism of systemic modernism' (Cooper & Burrell, 1988: 97). Bureaucracy is the characteristic organisation – of large government organisations, non-profit organisations and businesses. But the 'iron cage' of bureaucracy is not just a prison created by modernism. It is also a principle:

> As a principle it 'makes us free' to be modern. It makes us free because it is only through the purposefulness and goal directedness of organization that the uncertainties of disenchanted modernity could be coped with. (Clegg, 1990: 33)

Of course there are many empirical and theoretical nuances to bureaucracy. But empirically the pursuit of efficient control remains central. And theoretically, thinking about organisations remains largely the preserve of modernisation: rational action can lead to improvement and progress. The planning and calculating subject remains at the centre.

At the same time, there is growing disorganisation (Offe, 1985; Lash & Urry, 1987; Crook *et al.*, 1992). The complex, large-scale and monolithic organisation increasingly misbehaves when faced by the unpredictability and turbulence of the capitalist market and social developments. That is why current thinking about organisations is finding more and more inspiration in postmodernism and its attention to contingency, unpredictability and fragmentation (Bergquist, 1993; Burrell, 1988; Clegg, 1990; Cooper, 1989; Cooper & Burrell, 1988; Hassard & Parker, 1993; Jackson & Carter, 1992; Nooteboom, 1992).

This growing disorganisation should lead to post-Fordist organisations, learning organisations, liberated organisations (Peters, 1992) and a host of others. A typical postmodern organisation would look like this:

It would tend to be small or be located in small subunits of larger organizations; its technology is computerized; its division of labour informal and flexible; and its managerial structure is functionally decentralized, collective, and participative, overlapping in many ways with nonmanagerial functions ... (it) ... tend(s) to have a postbureaucratic control structure even though prebureaucratic elements such as clanlike personalism, informalism, and corporate culture may be used to integrate an otherwise loosely coupled, centrifugal system. (Heydebrand, cited in Clegg, 1990: 17)

But many writings about postmodern organisations have an idealistic ring to them (e.g. Clegg, 1990; Jackson & Carter, 1992). They are based on the conviction that the postmodern organisation is a better organisation, and, in particular, less bureaucratic. They are doubly wrong: on the one hand, they describe bureaucracy – a specific, historical pattern of organisation – in normative and usually pejorative terms; on the other hand, they apply postmodernism prescriptively. This is remarkable, considering that mistrust of grand narratives and a denial of progress are characteristic of postmodernism.

In a postmodern conception of organisations, much more attention should be paid to aspects like the textuality and intertextuality of organisations (see Van Twist, 1995; Cooper, 1989); the homogeneity and heterogeneity of forms of organisation (Burrell, 1988); differentiation in and between organisations (Cooper & Burrell, 1988); the undefined and paradoxical in organisations (Cooper & Burrell, 1988); pleasure and desire in organisations (Burrell, 1993); decentred subjectivity (Poster, 1990); and multirationality (Snellen, 1987).

The question is whether the terminology of organisation and organisations is acceptable, given their deep-seated association with modernism, modernity and modernisation. An attractive alternative image might be the 'rhizome' or rootstock (Deleuze & Guattari, 1977). The rhizome grows in all directions, even faster, it is said, than it dies off. It has no centre and no linearity, and is therefore a good metaphor for postmodern forms of organisation and for a postmodernist conception of organisation.

Die Autoren setzen diesen zentrierten Systemen nicht zentrierte Systemen entgegen, Netzwerke endlicher Automaten, in denen die Kommunikation zwischen beliebigen Nachbarn verläuft, und Stengel oder Kanäle nicht schon von vornherein existieren; wo alle Individuen miteinander vertauschbar und nur durch einen momentanen Zustand definiert sind, so daß lokale operationen sich koordinieren und sich das allgemeine Endergebnis unabhängig von einer zentralen Instanz synchronisiert (Deleuze & Guattari, 1977: 28).

This image is startlingly similar to the descriptive variant of the network metaphor for organisations. That is to say, the variant which does not prescribe the network as a good form of organisation based on distributed power, but merely describes it, as a contingent pattern of organisation.

Contingent Networks

Postmodernity or, better, the process of postmodernisation is distinguished by disorganisation. Hyperdifferentiation and fragmentation result in the ties within the complex, monocentred organisation – however complex and differentiated it might be – beginning to pinch. High priests of management consultancy like Peters (1992) recognise this and derive some fast-selling recipes from it. They proclaim disorganisation as a new and desirable pattern of organisation. There is nothing wrong with that, and if it works, all the better. But my intentions are less prescriptive. The process of postmodernisation in organisations can be broken down into the following facets.

Flexibility
Market turbulence, as a result of consumerism, the fragmentation of preferences based on taste, and intensified competition, make greater flexibility a necessity. Flexibility requires a smaller scale of operation because large-scale operations involve delay and inertia. This flexibility applies to technologies, organisational patterns and labour relations. The quantitative and qualitative increase in the capacity of ICT in particular, makes the application of technology more flexible and differentiated. It also reduces the number of necessary feed-back loops to higher levels because it links knowledge directly to operational implementation. Organisation patterns (see chapter 3 and Crook *et al.*, 1992: 181–184) become more flexible through a more strategic orientation, the application of practices such as 'just in time' and quality control, teamwork and decentralised management (Crook *et al.*, 1992: 184–188). Different labour relations contribute to flexibility because a clearer distinction is made between the core personnel of the business, and the increasing numbers of temporary staff, emergency staff, outside assistance and home-based staff. This is accompanied by functional flexibility through task integration, 'multiskilling' and localised responsibility (Crook *et al.*, 1992: 188–193).

Such flexibility makes organisations more temporary and more fluid.

Market instead of hierarchy
In the co-ordination of an organisation's activities, the importance of the market is growing while that of the hierarchy is in decline. The increasingly popular creed of the entrepreneur (e.g. intrapreneurship in organisations) is evidence of that. We see as well that through the breaking up of organisations, through privatisation, through 'outsourcing', the number of contractual ties in organisations is on the increase. Hierarchical co-ordination along for-

malised bureaucratic lines of authority suffers the odium of delay and inflexibility, but, of course, remains in place. Even in the rhetoric of classical bureaucratic organisations like central government, the market and contract enjoy a prominent place.

All this reduces the fluidity and temporary nature of organisation patterns. At the same time, insecurity and uncertainty are increasing, even though the organisation patterns are becoming less hierarchical and organisation relationships more horizontal.

Organisational boundaries
Organisational boundaries are becoming blurred in the postmodernisation process. On the one hand, that is a consequence of ICT, because it strengthens the importance of horizontal electronic networks. On the other hand, because of flexibility and market co-ordination, horizontal organisation relationships – both within and between organisations – are taking over from the vertical relationship between centre and base. The complexity of property and financial relationships further contributes to this picture of organisation patterns as communication and exchange relationships, rather than physical entities with a strong vertical integration. Codes of meaning in organisations are following more and more the temporary patterns of activity instead of an organisational tradition. The constant harping on about the importance of organisation culture is a – possibly regressive – reaction.

Knowledge intensity
ICT and other technologies have increased the knowledge intensity of production processes. Production relations are therefore distinguished by symbiotic configurations of humans and machines. On the one hand, this leads to a decline of the individual subject in organisations. On the other, it requires higher professional qualifications. Furthermore, professional autonomy is increasing as average qualifications rise. This has paradoxical consequences: a lessening of subjective autonomy through knowledge intensity and an enlargement of decentralised autonomy through further professionalisation. Central management's significance is becoming tragically mythical as its powers are decentralised. The significance of management as such is diminished by the intellectual capacities of technology. Whether we can speak at all of subject and object, or of organisations as instruments, has become the question.

Self-organisation
One often hears postmodern organisations movingly described as learning organisations that need self-organisation and self-steering. It sometimes seems as if the democracy of the sixties has returned to the world of the or-

ganisation. In more down to earth terms, we are talking about the consequences of postmodern fragmentation. Turbulence and fluidity, blurred boundaries, market relationships and knowledge intensity have turned centralised, panoptic steering into an illusion. There is no overview, knowledge is insufficient, unpredictability has become a principle. Self-steering and self-organisation are now the last resort for the modernist passion for control. If we accept that each notion of steering and organisation is primarily a cultural, and therefore a historical construction, we can then accept that steering and organisation are the contingent outcomes of intentions, passions, rationality and irrationality.

The rhizome metaphor powerfully represents the fact that postmodernisation means that organisations are contingent networks. Two other postmodern processes add to the power of the metaphor. They are both cause and effect of the facets of organisational postmodernisation which we have been describing. They are 'time–space compression' and 'economies of scope'.

Time–Space compression

The compression of time and space is one of the most important discontinuities in modernist traditions. In the place of an absolute and progressive interpretation of time and space, there has come a more relative and ephemeral conception (Mommaas, 1993: 99). According to Giddens (1990), the intension and extension of modernity is without precedent. Through the speed and extent of change, modernity, in an institutional sense, has become multidimensional. Capitalism and industrialism have spread world-wide, causing time to be separated from space, and subsequently space to be separated from place (Giddens, 1990: 17ff.). Giddens speaks of a 'distanciation' of time and space. Time, space and place are becoming independent of each other. Harvey, on the other hand, talks of the 'compression' of time and space in modernity. In both cases, there is a divergent movement of homogenisation and fragmentation. On the one hand, modernisation is without doubt a process of globalisation, and thereby of uniformity, standardisation and homogenisation; on the other hand, there is compression within fragmentation because local specificity is becoming more important: the free movement of capital around the world increases the importance of place as its specific characteristics are used to attract capital.

Time and space, according to Giddens and Harvey, are becoming permanently restructured. By separating them, and subsequently recombining them, they are being socially and economically manipulated. Social relationships are being less specifically and locally determined; they are taking shape along other dimensions. Knowledge and information lead to an awareness of the permanent, though usually ungovernable, changeability of modernity. (See Mommaas, 1993: 105–106)

At the same time, the national state constitutes a counter-movement. There, time and space are tied together: in a territory, in a national culture, in a specific political time-scale. Instead of the 'privileging of time above space', we see in this aspect of modernity a restoration of linearity and the centre, and therefore a privileging of space above time (Harvey, 1989: 271–273).

The process of postmodernisation moves in the opposite direction. Over and against the relative homogenisation and linearity of modernisation, we see the process of hyperdifferentiation and fragmentation. 'Being' becomes more important than 'becoming'. The importance of place increases because a specific identity can be attached to it. Place, however, is not necessarily territorially defined. It can be a virtual community on a world-wide scale. At the same time, the importance of space and time are reduced, because what counts now, are fragments, specific combinations of styles, which are, however, as ephemeral as they are changeable. Style, the community, identity are tied neither to space nor to time. Hence, postmodernism plays down the notion of a physical organisation with a place, a tradition, and an ordered time-scale.

The temporary and the ephemeral take the place of ordered and organised time; fragmentation and de-territorialisation replace ordered and organised space. Thereby the cohesion of time and space disappears. The importance of scale diminishes, while that of reach, of scope, increases.

Economies of Scope: from Scale to Scope

As time and space become more relative – especially through developments in ICT, distribution and organisation – as the economic importance of services increases at the expense of goods, and as taste and style form the primary impetus of consumption, it becomes more and more important for organisations to consider 'economies of scope'. How can a product be made as specific as possible for a specialised niche in the market, and sold as quickly as possible throughout the world? It is no longer enough to produce as much as possible of the same product: economies of scale have become less important as a gauge of efficiency. Added to that is the fact that, because of developments in technological capacity and its increased knowledge intensity, the small-scale has, for the first time in history, become a technological option. That means that as varied a package of goods and services as possible best responds to the fragmentation of preferences and markets. But the changeability and ephemerality associated with this fragmentation make specialisation absolutely essential. Organisations must focus on very specific products and even on very fragmented clients and markets. That again encourages both further fragmentation in organisations as entities, and continual changes in network patterns for specific economic activities.

The scope of organisations, not just in a quantitative sense but primarily in a qualitative sense, is becoming a more important criterion for survival than their scale. In fact, the large-scale conflicts with the processes of postmodernisation. Thus an advertisement for a computer company refers rhetorically to the extinction of the dinosaurs as an evolutionary necessity.

Once again, fashion offers a striking illustration. The distinction between *haute couture* and the high street is vanishing. The consumer no longer purchases a single collection for the season. No longer is there a clear-cut 'line' for each season. Fashions are now much more emphatically defined by the lifestyles of mass culture. Eclecticism is rife; indeed, it is the only consistency left to fashion. Changeability is so great that retailing and 'sales' are indistinguishable. But because of this changeability, clothes become outdated much more slowly. The basic outfit is completed by special offers from a range of styles and collections. This is seriously threatening the fashion industry, even though the sector has long been familiar with a turbulent market. Postmodernisation means that the industry can no longer dictate the market. It is now done by the consumer, and the identity-defining actors from other sectors such as music, youth culture, sport and the mass media.

We can see from this that concentrating on 'economies of scope' is no simple panacea. Scope becomes increasingly difficult to define because of postmodern fragmentation. Furthermore, scope is changeable and ephemeral: it withdraws from linear developments and can not even be pinned down as a plurality of centres. Postmodernisation is a process of decentring.

7.6 AMBIGUITIES

This discussion of postmodernity, postmodernisation and postmodernism is neither complete nor unambiguous. Nor does it have to be. Firstly because it provides a background for the following chapters. It is an intermezzo that links together the first six chapters with some theoretical ideas. Thereby, patterns and perspectives have become visible to form an overture for the final three chapters. Secondly, because completeness and unambiguity are hopeless as well as dull ambitions. Hopeless, because a narrative about fragmentation is itself a fragment; dull, because ambiguity, vagueness and lack of clarity are more exciting. Moreover, they are non-totalising.

To claim, against this background, that the age of postmodernity has arrived would be nonsensical and impossible. Nonsense because the narrative would merely contain the obvious and the self-evident. Furthermore, we can not even be sure that modernity has yet been completed in all respects. According to authors such as Beck, Giddens and Lash (1994), one should rather speak of a state of reflexive modernity. Other writers, such as Habermas (e.g. 1985b), argue that modernity is an incomplete project to which intellectuals at least should lend their best efforts.

'Postmodernisation' is therefore more attractive than 'postmodernity'. It holds open the possibility of continuity with modernisation while simultaneously indicating fractured surfaces and new tendencies. The process of postmodernisation is multi-dimensional and potentially as far-reaching as modernisation, of which it is a continuation but whose stability it undermines. Thereby, transformation occurs. Differentiation, commodification and rationalisation become disorganisation, hyper-differentiation / de-differentiation, hyper-commodification and hyper-rationalisation. The crucial element is the tension in the modernisation process between bureaucratising tendencies and hyper-differentiation. The changes which result from this are uncertain and unpredictable. The core of this is supplied by the ever more penetrating role and significance of the cultural factor (Crook *et al.*, 1992: 220). A sociological analysis of postmodernisation could be directed, intellectually and pragmatically, as follows:

- to identify the emerging/shifting foci of uncertainty in the processes of postmodernisation;
- to study the ways in which principles of uncertainty in any register become implicated in social action;
- to alert publics and decision makers to the rich possibilities for irony and reversal in any attempts to plan and manage change. (Crook *et al.*, 1992: 238).

And although the last point sounds somewhat pious, it contains within it – the possibilities for irony and reversal – the most attractive theoretical elements of postmodernism. Postmodernism is ironic – however difficult that may be – and desires passionately to avoid the lure of metanarratives and metaphysics. That passion is based on disbelief, and a fear of the totalising nature of metanarratives and metaphysics. Plurality, contingency and uncertainty are not insoluble puzzles, but attractive aspects of postmodernity which, in postmodernism, should be honoured.

> There are many different word-games – therein lies the heterogeneity of its elements. Only in sections do they lead to institutionalisation – therein lies local determinism. (Lyotard, 1987a: 26–27)

Focusing attention on what is attractive is very important. For when we become serious, there is a danger that we start to cherish the illusion 'that there is a higher political goal than the avoidance of cruelty' (Rorty, 1989). A constant search for contingencies, uncertainties, vagueness and ambiguity is the best guarantee of avoiding that danger. I can do no better than to finish by again quoting Rorty:

> My essays should be read as examples of what a group of contemporary Italian philosophers have called 'weak thought' – philosophical reflection which does not attempt a radical criticism of contemporary culture, does not attempt to refound or remotivate it, but simply assembles reminders and suggests some interesting possibilities. (Rorty, 1991b: 6)

8 Administration and Politics in Postmodern Cyberspace

In the theoretical intermezzo, I dealt with postmodernity, postmodernisation and postmodernism in general terms. More specific attention was given to the postmodern view of organisations and organising because of its relevance to this chapter.

I shall now return to the central themes of my narrative: politics, administration and technology. What is the significance of postmodernisation and postmodernism for these themes? (I have already pointed out that it is unhelpful to speak of 'postmodernity'.) Are processes of postmodernisation taking place in the worlds of politics, administration and technology? And what value might a postmodernist perspective have for them? These are the questions which I shall address in this chapter.

Before doing so, one or two observations are in order. The technological theme may appear a little curious. But many postmodern writers consider that developments in the field of ICT are the driving force behind the process of postmodernisation and that the specific effects of ICT contribute to a postmodernist perspective. Lyotard and Baudrillard are especially conscious of the consequences of the massive expansion of ICT.

> It is reasonable to assume that the growth in the number of information-machines is affecting, and will affect, the spread of knowledge, in the same way as the development of means of transport: first people (vehicles) and later sounds and images (media). (Lyotard, 1987a: 31)

Baudrillard in particular sees ICT as one of the most important forces in postmodernising our reality. Technology produces images, simulacra which have classically been taken as representations of an 'objective reality'. But images have now become reality itself: hyper-reality.

> Today, the entire system is fluctuating in indeterminacy, all of reality absorbed by the hyperreality of the code and the simulation. It is now a principle of simulations, and not of reality, that regulates social life. The finalities have disappeared: we are now engendered by models. (Baudrillard, 1988: 120)

A postmodern interpretation of the information society is therefore in a sense tautological. In the literature of postmodernism, ICT is regarded not just as a trailblazer of postmodernisation but as a direct producer of the postmodern experience: of postmodernism.

8.1 TECHNOCULTURE

The radicalism of ICT lies in its intellectual character: ICT is the codification of theoretical knowledge (Bell, 1979) and also a substitute for the subject as its primary bearer. (See also Coolen, 1992.) That is the core of what Poster (1990) calls the 'mode of information'. Because of its cultural significance, I use the term 'technoculture' to indicate that cultural actions – informing, communicating, image-creating – are the primary objective of the technology, and to place changes in codes of meaning at the centre of the analysis. More than ever before, we live in a technological culture.

Decentring the Subject

Modernity is the era of decisionism. Reality exists and moves because we take decisions. The human subject is the measure and the motor of everything. In this, modernism's desire for emancipation harmonises with the cybernetic obsession with control. Optimal freedom is optimal control, and the course of history is proceeding inexorably towards the successful realisation of that combination, or it should be steered in that direction. After all, modernity is an ongoing project (Habermas, 1985b).

ICT marks the transition to another era, to another culture. The subject is becoming decentred – a process which is observable at various levels. In the first place, technology is now competing seriously with the subject. (See Turkle, 1984: 306ff.) The problem, as Poster puts it, is that we are 'becoming the subjects of the instruments we created to be our subjects' (Poster, 1990: 148). In a number of respects, man is becoming subordinate to the machines which he has created. Those machines are no longer just an extension of our physical capacities; they are also an extension of our intellectual capacities. McLuhan declared that people have henceforth externalised their brains and sinews. While ostensibly he was only writing about the mass media, this view has in retrospect taken on a particularly prophetic significance (McLuhan, cited in Crook *et al.*, 1992: 67). And even though artificial intelligence has for decades been greeted with justifiable scepticism, its development seem likely to accelerate exponentially. (See e.g. Kelly, 1994; Mieras, 1994c; Levy, 1992; Negroponte, 1995.)

Secondly, there is ICT's effect on subjectivity. Subjects are constituted by and through communicative actions and structures via computer networks. Technology thereby changes the individual's relationship with the world (Poster, 1990:11). Through those actions, the subject is decentred, spread and multiplied in and between innumerable networks and databases (Poster, 1990: 6–7). No longer can one speak of stability, or at least the stability of the subject as the centre of the world. And that applies even more where the relationships of the subject with different areas of social activity – administration, politics, economics, work, leisure, and even love – are electronic relationships. Information breaks free of the carrier, and, outside his influence, becomes plentiful, available to many, and difficult to commodify. It is not for nothing that ownership, in a legal, economic and subjective sense, has become a thorny problem in the information society. And identity is a form of fiction in the structures of communication (Poster, 1990: 117).

In ICT, codes of meaning that no longer have the subject as source and centre are institutionalised. At the same time this institutionalisation is transient and changeable and distinguished by permanent instability. Technoculture is a staccato culture (Zijderveld, 1991).

Images, Images, Images

Technoculture is also an image culture. This is true at the level of opposition between words and images, but even more so at the level of representation. Ever since the expansion of mass media (especially television) which was strengthened by the globalisation of the entertainment industry, the cultural advance of the image as the bearer of culture has been unstoppable. Television represents and creates reality, while the identity of children is being increasingly influenced by computer games. Though early varieties of ICT were strongly text-oriented, the impressive improvement in graphical software and visual communication has given a powerful impetus to its expressive and representational character. The integration of classical media and ICT, which is surely imminent, will give a massive boost to the process because practically every household in the modern world will be absorbed into the networks of the image culture.

Another feature of the image culture is even more far-reaching, if that is possible. Images, it is still argued, are representations of reality. So ICT distributes representations of reality, whether in the simulated reality of computer games or through images of economic and social developments based on computer-generated models and scenarios. However, because ICT has expanded so enormously – from the private world to global networks – these representations are taking on an autonomous reality. Communication now consists in the interchange of representations of representations of representations ... Representations are thereby becoming self-referential: the hyperreality of simulacra, in Baudrillard's words. Postmodern anti-representationalism, as formulated for instance by Rorty, corresponds to the 'empirical' reality of ICT-generated images. The widely-held belief that 'the computer makes no mistakes' is a prosaic illustration. The screens which banks present to their customers appear to have broken through the incomprehensible magic performed by bank clerks behind their computer. But in fact, their magic is even more powerful: the screen has become reality.

The image culture of postmodern representations is inevitable because of the quantitative expansion of ICT, the qualitative improvement and potential of ICT, and through the increasing substitution and support of practices and transactions by ICT.

And the image culture is endlessly fragmented not just because it represents a fragmented reality, but primarily because the representations themselves create a fragmented reality with autonomous significance. In a discussion of science and the postmodern, Poster expressed it as follows:

> The electrification of science in the form of the computer generates a discourse at the border of mind and matter and in so doing destabilizes the distinction between the two. (Poster, 1990: 149)

The Internet as Metaphor

The Internet – the world-wide network of computer networks – is one of the most sensational applications of ICT and a splendid metaphor for our technoculture. The Internet is a loosely linked network of networks which have communications technology at their core. Via the Internet, it is possible to distribute texts, images and sound. The Internet has no central organisation, but is primarily a set of agreements about communication standards. Somewhat ironically, it was originally thought up at the height of the Cold War by the American military-industrial establishment as a communications infrastructure which, just because it had no centre and consisted only of connections, could not be destroyed by military attack. Since being opened to the universities, its growth has been explosive. There has been a colourful proliferation of activities on the Internet: discussion lists, discussion groups, communities, scientific communication, presentations by organisations, institutions and businesses, buying, selling, trading, political conspiracy, sex and drugs and rock and roll, love affairs, games, e-mail, meetings, correspondence, writing, art and so on and so forth. Read *Wired*, the American magazine which deals with many facets of the technoculture, for a more detailed impression. (See also Rheingold, 1994.)

The Internet is a metaphor for postmodern technoculture for a variety of reasons:

- It is anarchistic because it has no central authority or rules.
- It is self-regulating, because the users define norms and rules for their own communities, and the 'exit' option is always available.
- It is fragmented, because there is no linear process of differentiation, it proliferates and expands without any clear logic: it is an example of the rhizome (Deleuze & Guattari, 1977).
- It decentres not only itself but also its users, because their identity is drawn from the different communities and activities on the Internet which possess no stable unity.
- It is independent of time and space and so creates non-territorial communities.
- It produces a reality of representations which are self-referential.

The metaphorical significance of the Internet is so powerful because its specific technical form and its social, political and economic potential are so attractive that a growing number of communication infrastructures will take on an Internet character. Whatever efforts are made to keep some infrastructures closed and secure, however ambitious the plans are for regulation and uniformity, the spread of ICT on a world-wide scale into every single household makes central control and domination an illusion. That is not to say that control and domination are disappearing, but that they too are multiplying and fragmenting. Even the electronic variant of Foucault's microcosms of power will not produce an unambiguous and universal panoptic view.

Cyberpunk and Big Brothers

Technoculture can not be described as a new utopia, any more than postmodernisation can have ambitions for a better world. Indeed, because postmodernism and technoculture are so intertwined, we should be extremely sceptical about any attempts to present Technoculture as a new grand narrative of enlightenment and progress.

Technoculture is equally attractive to large financial conglomerates and to anarchistic hackers; it is as attractive to Newt Gingrich as it is to The Grateful Dead, to right-extremist bombers as to the security forces, to the idealists of the Digital City as to direct marketers, to the Chairman of IBM as to the burgomaster of a tiny township. In political and moral terms, the technoculture is empty because it involves representations which do not claim any objective truth. Technoculture expresses the anarchism of the market on a global scale. It is precisely that combination of market and anarchism which gives to technoculture its postmodern identity of non-directedness, ephemerality and changeability.

An alternative name for technoculture is 'cyberpunk'. It refers to a cultural, political and social orientation, a vision of the future in cyberspace. Most of its ideas are borrowed largely from science-fiction, films, computer games and innumerable 'virtual' communities. It indicates a reality of subcultures and 'designer cults' with their own codes and lifestyles, where people and computers form symbiotic unities. (See Burrows, 1995: 3ff.) In a social and political sense, a powerful discontinuity can be observed from the stable connections, centrally organised movements, the clear political contrasts, and the wide-scale bureaucracies of the modern capitalist world. Cyberpunk is anarchistic in practice rather than on principle.

Over and against that, of course, we have the world of Big Brother which is being materialised in ICT. As I have shown in earlier chapters, ICT embodies the modernist desire for perfection, rationality and domination. Infrastructures and architectures for information provision and communication are the new virtual strongholds which are replacing classical bureaucracy. (See Zuurmond, 1994.) The infocracy has no intention of abandoning its ambition to dominate and control: on the contrary, it is substituting the rigidities and obstacles of classical bureaucracy by the flexibility, speed and transparency of ICT.

But there are many Big Brothers, all in fierce competition with each other, and all hitting the limitations of monocentric organisation. The small-scale power of ICT means that cyberpunk is always partly present as a counterweight, not so much against as within the Big Brothers. Both, after all, are part of the same fragmenting and decentring technoculture that is more archipelago than pyramid. This does not necessarily make the archipelago any more peaceful or friendly than the pyramid.

8.2 VIRTUALISATION

Anyone looking for a splendid description of virtual reality should read *Disclosure* by Michael Crichton (1994: 392–402). As so often, a novel can recreate a specific reality much more vividly than a dry technical treatise. Extensive surveys of the current state of affairs - though this can change at any moment – as well as of the social, cultural and organisational consequences of applying technology can be found in Rheingold (1991 and 1994), Sherman and Judkins (1992) and Schroeder (1993). Virtual reality is a specific application, a configuration of different technologies. I described them in Chapter 3. I have used virtualisation as a metaphor for the significance that the expansion of ICT has for organisations, politics and administration. In academic writing, virtual reality is discussed in much the same way: on the one hand, as a particular, technological application; on the other, as a new reality in cyberspace.

> Cyberspace is a physically inhabitable, electronically generated alternate reality, entered by means of direct links to the brain – that is, it is inhabited by refigured human 'persons' separated from their physical bodies, which are parked in 'normal' space (…) To some extent, though, cyberspace already exists as a metaphor for late twentieth century communications technologies (databanks, financial systems, ATMs). Many of us already live at least part-time in cyberspace, but call it 'computer conferencing', 'phone sex', banking, or 'virtual' this or that. (Stone, 1992: 609–610)

It is precisely this virtualisation that contributes in an important way to the processes of postmodernisation and whose effects on understanding and knowledge have inspired postmodernism as a theoretical position.

Time–Space Compression

In general, the application of VR and ICT provides organisations with all kinds of opportunities to construct and reconstruct reality. Information is processed very directly so that immediate reflexive information about numerous activities can be produced. The Air Miles scheme represents in a simple and fundamental fashion the 'informating' capacities of ICT (Zuboff, 1988). Evaluation can be continuous, and complicated, time-consuming feed-back processes, involving functionally differentiated parts of the organisation and top management, are unnecessary. Organisational and policy-oriented information does not have to be kept separate, either temporally or spatially. Policy-relevant information about implementation is immediately – that is to say, non-sequentially – accessible, as large-scale transaction systems become equipped with intelligent applications to keep track of patterns, marginal values, critical factors, profiles etc. within the transactional data. That is precisely the point of cash points, plastic money and chipcards. Each electronic connection between citizen and government generates information which can be used in policy-making. When it involves processing information, time as a factor in organisational and policy processes is compressed. Information and meta-information is instantly available at any given – and often unwanted – moment.

There is also spatial compression because no separate organisational subdivisions have to be created for collecting information. Through the quantitative and qualitative growth in the capacity of systems, networks and applications, implementation and development of policy can occur within the same organisational units. And given the need for flexibility and fast response, the obvious thing to do is to exploit the links between development and implementation within differentiated small-scale implementing-organisations. In the light of actual practice, the widely trumpeted distinction between policy and implementation never made a lot of sense; in the perspective of ICT developments it is becoming complete nonsense. The time–space compression created by ICT also leads to a compression of the policy process and to 'functional compression' within organisations.

Because of time–space compression, time and space also become less relevant as factors in policy-making and organisation. It is just as easy to physically separate organisational functions and phases in the policy process as it is to integrate them in the first place. Information collection, distribution and processing can occur in geographically scattered locations and by differentiated organisations. Physical integration of these activities is no longer necessary since integration can take place over electronic networks. This is how virtual organisations arise which are held together not by a vertically integrated bureaucracy but by a complex pattern of contractual connections and electronic infrastructures. The data-stores used by the organisation are themselves 'virtual' because at any given moment they can be reorganised on the basis of other data-stores. They appear only on a screen.

As a result, the boundaries between organisations become blurred. An activity is defined less by the structural configurations which we recognise as organisations and more by the infrastructures and architectures of information and communication. Horizontal patterns of information exchange and production displace the vertical command and control structures of the organisation as a machine or instrument of control.

Places without Territory: the 'Glocal'

> D'une part, le temps réel l'emporte sur l'espace réel; disqualifiant les distances et l'étendue au profit de la durée, une durée infinitésimale. D'autre part, le temps mondial du multimédia, du cyberespace, domine les temps locaux de l'activité immédiate des villes, des quartiers. Au point que l'on parle de remplacer le terme 'global' par 'glocal', une contraction de global et de local. On considère que le local est forcément global, et le global forcément local. Une telle déconstruction du rapport au monde ne sera pas sans effet sur la rélation de citoyen à citoyen. (Virilio, 1995: 28)

It is the classic image of the 'global village': the world becomes a village, the village a world. Social, economic and cultural activities and identities are no longer tied to a specific territory. The Internet creates countless virtual communities (Rheingold, 1994) in which 'real world' events can happen, but without their limitations of time and space. Places without territory are created. In them, activities take place and identities are formed. Although they are virtual in that they have no physical substratum (Snellen, 1994: 287), they have far-reaching consequences, as we can see in the transactions of virtual finance. According to Snellen (1994: 287), this virtualisation has various outcomes: it integrates different activities within a 'mental frame', it links successive steps of a process in a sequential framework; it creates a sense of geographical unity between territorially scattered processes.

Virtualisation is therefore a process of de-territorialisation. The spatial location of activities has become irrelevant in the sense that they can be relocated at will. On the one hand, the criterion of 'economies of scope' can be applied. Activities are carried out wherever is most efficient or effective: near a local market, where labour costs are low, in a suitable administrative environment. On the other hand, derivative criteria can also be applied: a pleasant neighbourhood, attractive educational or leisure facilities. The necessary connections can, after all, be established through ICT; physical proximity is no longer necessary. At the same time, the command and control potential of that same ICT remain intact and, indeed, can even be extended. The capacity for surveillance and monitoring is immense. (See also Dickson, 1990: 107–110.) That too is virtualisation: the boss no longer walks around or keeps an eye on things in the panoptic space of the Fordist organisation; he has become invisible in the information architecture and infrastructure of the infocracy.

> So a *virtual fortress* takes the place of Weber's 'iron cage' with its hard, rigid structure and palpable bars. Where the structures of the organisation were once focused on what its members did or did not do, informatisation now also focuses on what the organisation's members do or do not think. The rigidity of the iron cage is removed and the illusion of freedom is revived. However, the informational architecture invisibly determines and limits what may be thought, and thereby imposes limits on what may be done. Where the mechanistic order was limited in its power to impose complete obedience, the virtual order appears to be breathing new life into obedience as a normative standard. The categories in which it is permitted to think are laid down centrally, after which every member of the organisation is free to move around at will within those categories. (Zuurmond, 1994: 305.)

De-territorialisation is a form of decentring and one of the distinguishing features of postmodernisation. Whereas Zuurmond emphasises the uniformity and centralising character of the infocracy, it is more probable that a host of infocracies exist, without hierarchical order, and entangled in patterns of conflict and co-operation. And it is doubtful, even within a single infocracy, whether a centralised hold on the definition of reality can be sustained. For as the boundaries between organisation and environment become blurred, rival concepts, alternative narratives, can enter the organisation. Furthermore, the local systems in the organisation meanwhile become so powerful that they counteract Zuurmond's implicit image of the large mainframe surrounded by passive workstations. In the end, the ephemeral and mutable will prevail over solidity and stability.

Immediacy and Seduction

Cyberspace's technoculture is a virtual reality which creates an intense feeling of immediacy. Without obstacles of time and space, unbridled communication is possible and any need for information can be satisfied immediately and 'on line'. Admittedly this is rhetoric, but rhetoric is crucial in any culture. Complete information and unfettered communication are the myths of the 'information cult' (Roszak, 1986).

> With the advent of fast personal computers, digital television and high bandwidth cable and radio-frequency networks, so-called post-industrial societies stand ready for a yet deeper voyage into the 'permanently ephemeral'. (Benedikt, 1991: 11)

The sense of immediacy produced by virtualisation strengthens the fragmentation of identity because new and alternate codes of meaning are continuously being created. In this there is a high level of contingency just because of the immediacy of information and communication. An interesting development in this respect is that of 'softbots': intelligent software which registers the search behaviour of an Internet user, recognises patterns and, building on those patterns, carries out searches independently. No longer is it just tradition which generates patterns of meaning but computer software.

The fragmentation of identities and activities leads to ephemerality. The images, and therefore values and norms, change at such a fast tempo that there is no possibility of anything sticking permanently. ICT is a powerful stimulus in this form of postmodernisation. The ephemeral is no longer some negative side-effect of the modernising process, but is becoming a central feature.

> Terrified of being alone, yet afraid of intimacy, we experience widespread feelings of emptiness, of disconnection, of the unreality of self. And here the computer, a companion without emotional demands, offers a compromise. You can be a loner, but never alone. You can interact, but need never feel vulnerable to another person. (Turkle, 1984: 307.)

Next to the cultural significance of the immediacy produced by ICT, there is an organisational impact. Various writers talk about transparency: the absolute visibility of the activities and processes which ICT generates. But it is not only visibility in a direct sense (enabling unprecedented panoptic control), but it is also analytic and intellectual insight (Snellen, 1994: 287). The 'informating' capacity of ICT, as Zuboff puts it, makes it possible to generate immediate knowledge about processes: patterns and profiles. These patterns and profiles can subsequently become the input for new activities and processes, such as strategies of control.

In other words, ICT offers a very seductive method of achieving the classical ambitions of the modernising process: rationalisation and control. The boundaries of understanding and knowledge appear to be transcended by the inexhaustible supply of information: 'more knowledge, better decisions' according to the publicity of the computer division. The limitations and rigidities of classical bureaucracy are removed. The post-Fordist or postbureaucratic organisation is flexible, fast and fully informed. Hierarchy as a mechanism for co-ordination has been replaced by the architecture of information provision.

> Information systems that translate, record, and display human behavior can provide the computer-age version of universal transparancy with a degree of illumination that would have exceeded even Bentham's most outlandish fantasies. Such systems can become information panopticons that, freed from the constraints of space and time, do not depend upon the physical arrangement of buildings or the laborious record keeping of industrial administration. They do not require the mutual presence of objects of observation. They do not even require the presence of an observer. (Zuboff, 1988: 322.)

And because control and domination are exercised with the help of ICT activities and processes, the informational panopticon becomes a reality of images and representations: postmodern virtualisation.

Administration and Politics

Postmodern technoculture, with its virtualising of subjects, time and space, organisations, territory, and activities, has important consequences for administration and politics. In the rest of this and succeeding chapters they will be dealt with at length. Furthermore, they will be looked at in relation to other administrative and political processes which are not directly linked with technological developments. To conclude this section on technoculture and virtualisation, I shall summarise what some of the consequences for politics and administration might be.

- Decentring the subject affects the perception that administration and politics are real products of decisions and deciders. But this perception that social developments are the intended result of decisions made by subjects and institutions has been weakened because ICT has become a significant competitor. It affects administration and politics in particular because they have always been regarded as the decision-

making machine *par excellence* in matters of administration, policy and ideology.
- The decentring of the subject is being advanced by the administration through its widespread use of ICT. Public administration thereby contributes unintentionally to social fragmentation.
- The image culture of ICT is competing with the administration's definitions of reality and the grand narratives of politics. The strong emphasis on representations, and representations of representations, undermines the claims of political ideologies to truth and reduces their utopian character. The result is fragmentation.
- The characteristics which the electronic reality of systems, databases and networks have borrowed from the Internet – anarchism, self-regulation, fragmentation, decentring, virtualisation and self-referentiality – conflict with the institutional characteristics of politics and administration: order, regulation, coherence, centralisation, territoriality and intervention.
- Cyberpunk indicates a colourful collection of lifestyles which agree in their rejection of classical politics and modern administration. It is a form of pragmatic anarchism that, precisely because of its pragmatic character, has withdrawn from traditional political and administrative cultures.
- Public administration is undoubtedly one of the Big Brothers. On the one hand, there is increased competition from other, often globally-operating, Big Brothers. On the other hand, the nature of the technology on which the Big Brothers are based makes it probable that the social archipelago will expand further. That undermines any kind of pyramid, including the pyramid of politics and administration.
- Time–space compression has fundamental consequences for the structure and functioning of public administration. More and more, policy processes are operating interdependently and in parallel instead of in a traditional sequential fashion. The structure of organisations can, on the one hand, become more differentiated and small-scale while on the other hand, there is increasing functional integration within and between units at a horizontal level. Classical bureaucracy is thereby becoming obsolete and political primacy is being adversely affected.
- De-territorialisation undermines the notion of the nation state as representing the political and administrative integrity of a specific region. That mainly affects political institutions.
- The immediacy that ICT produces competes with the traditional mediating and deliberative role of politics. The resulting fragmentation can no longer be cured by the consistency of a grand narrative.

- ICT is also seductive as a means of extending the modern state's ambitions for control and discipline. However, the normative foundation of those ambitions is being eroded.

Technoculture and virtualisation do not make politics and administration disappear. However, their shape is changing fundamentally. According to my observations, the process of postmodernisation is leading to the rise of the virtual state. To this we shall now turn.

8.3 GOVERNANCE: INTELLIGENCE AND AESTHETICS

In the study of public administration, the themes of postmodernisation and postmodernism have scarcely been discovered. Here and there, we come across scattered references: Abma (1994), Boutellier (1994), Hummel (1990), Kuypers (1994), Noordegraaf (1995), Van der Loo and Idenburg (1994). More systematic treatment can be found in Crook *et al.* (1992), Reiner (1992), Slaats and Knip (1992). Finally, there is the recent study by Van Twist (1995) which is not explicitly postmodern in its approach but can be read in that light, Willke (1992) who attempts to formulate an ironic constitutionalism – an ambition which I share – and Fox and Miller (1997) who stress the narrative-oriented approach. The question to be addressed now is what significance postmodernisation and postmodernism have for public administration. To answer this I shall refer back to the developments sketched in Chapters 2 and 4, while my interpretation is built on the perspective formulated in the theoretical intermezzo.

Intelligent Fragmentation

Public administration is a collection of bureaucratic organisations under political direction within a framework of party-politically structured democratic responsibility. Bureaucracies are the chief embodiment of modernisation's rationalising tendencies and ambition to dominate. The vicious circles of bureaucracy (Crozier, 1963; Vroom, 1980) symbolise its rigidities. They arise from an inherent contradiction between processes of differentiation and hyperdifferentiation peculiar to modernisation on the one hand, and the artefact of monocentric organisation on the other. This inherent contradiction carries with it a process of 'paradoxical modernisation'. There is a permanent urge to modernise the bureaucracy: numerous major or minor reorganisations; the introduction of business methods and attitudes; attempts to make the instruments of policy and control more sophisticated. The paradox lies in the fact that although bureaucracy is being further rationalised, it is only happening through emphasising a single dimension of rationality. The resulting clash of rationalities (Snellen, 1987) is becoming increasingly obvious.

The rationalisation of bureaucracy should contribute to the effectiveness and efficiency of the monocentric organisation. Hence, achievement indicators are popular criteria for control. (How different is this from Stalin's Five Year Plans?) At the same time, we can observe processes of differentiation and hyperdifferentiation. I have mentioned three: localisation, functionalisation and autonomisation. The case of independent agencies bore this out. As a result of these processes, monocentric organisation becomes relatively less important. Responsibilities are distributed throughout a complex of autonomous and semi-autonomous units. Tasks become differentiated among various organisational departments which are specified by profession or divided according to function. Privatisation leads to another division, that between the private and public sectors, though the demarcation between them is blurred by the expansion of public–private arrangements. Crook *et al.* add internationalisation to the factors which contribute to differentiation (Crook *et al.*, 1992: 101–102).

An important underlying principle for this differentiation is that the structure and functions of public administration should reflect the pluralising of social domains. The shortcomings of monocentric organisation are countered by differentiating public administration in both its structure and its practices. The pattern is comparable to the process of flexible specialisation in the business world (Harvey, 1989; Clegg, 1990). It is theoretically inspired by configuration- or network-theory and the new *autopoiese* or social systems theory (In 't Veld, Schaap, Termeer & Van Twist, 1991).

However, it is inevitable that the process of differentiation, which from a modernist perspective might still be seen as arising from a centre and moving linearly, will result in hyperdifferentiation. Steering, to be effective, now requires such a wide variety of structures and practices, such adaptability to social change, and such rapidity of intervention, that notions of centre and linearity are becoming obsolete. The effect of hyperdifferentiation brings about a number of reverses which we can designate as fragmentation.

The pyramid form of public administration is being undermined and changed into an archipelago of relatively autonomous units over which central political and bureaucratic authority can only be exercised with the greatest difficulty: it increasingly takes the form of rhetoric, of triviality, or of a blueprint which excludes variety. Substantive policy-making is carried out in the decentralised implementing organisations and their networks of public and private sectors (see Huigen, 1994). This makes the outcome of policy-making contingent and organisational structures highly variable. More and more, contractual links between relatively autonomous units are replacing hierarchical co-ordination through decree, command and control.

The form of the administration thereby becomes accidental: not the planned outcome of designers but the fragmented result of social developments and intelligent reactions to them. From a monocentric point of view, the fragmentation of public administration is problematic. But in the light of the postmodernising processes described in Chapters 6 and 7, fragmentation is an intelligent response. Although fragmentation is usually fortuitous, public administrative organisations can attempt to use its accidental nature intelligently as a basic principle in its design.

Agreeable Co-production

There have been significant changes in the practices of public administration, parallel to and linked with the process of organisational fragmentation. The perception that social domains have a network character and consist of many actors who are mutually autonomous (in terms of authority) and interdependent (in terms of strategic needs) is being reflected in the design of steering and policy-making. Policy processes are either organised on the basis of this network character or linked to it. The government is thus either just one of the actors in the network or its director. Sometimes, of course, these roles can become confused.

Within the networks, coalitions of actors are sought who will make policy-making more effective. Institutionalising the network can also help to create more or less stable policy-making practices. However, changes in such arrangements are greater than was usual in the Dutch neo-corporative tradition. The players change more frequently, the rules of the game become the object of debate and negotiation, outcomes are contingent. Public administration thus focuses more on outcomes than on the design of the arrangements. Steering becomes intrinsically more abstract and procedural. (See Frissen in Kuypers, Foqué & Frissen, 1993.)

Policy-making may be regarded as a form of co-production. Different actors are involved and make contributions. Actors may be chosen in the process of designing the arrangements, but much more often, most will demand a role for themselves. The power of the many single-issue movements is enough to ensure that.

Consequently, any assessment of the effectiveness of policy-making acquires an aesthetic rather than a politico-ideological character. The importance of support, consensus and acceptance means that greater weight is given to skills in organising successful co-production in public administration. The design of co-production is also critical: which parties, which mechanisms, which processes, which intermediaries are questions which all must be addressed. The quality of the policy-process is not measured by its actual results but by its design and by the degree to which it is based on consensus and participation. Those are aesthetic norms. Co-production is pragmatic because practicability and feasibility are put first. Co-production is agreeable because involvement and consensus are its aims, and conflict is avoided or regulated by mechanisms for conflict-management.

Contingent Steering: Images and Infrastructures

Any steering which takes place in these fragmented networks is naturally highly contingent. Many of the actors, including administrative organisations, now deliberately avoid the classical model of governance. That model is top-down, assumes a clear hierarchy, a centre and linearity in policy-making processes, it postulates the policy-makers' monopoly over the policy object, and it sees policy as the mechanistic implementation of a blueprint. If classical steering is to be effective, it requires full information about alternatives, familiarity with effects and side-effects, a validated policy theory, predictability of, and control over, social developments. Admittedly this is a caricature. No-one will ever have actually believed in the empirical validity of this model. But as a lodestone, a guide-line, a utopia, it has been a powerful inspiration. Even today, the design of organisations still bears its imprint.

In new forms of steering (see Bekkers, 1993), policy makers now appear to be resigned to the contingencies of steering in a fragmented society. This has led to attempts to use fragmentation intelligently as a basic principle since the plurality of social conditions requires a variety of administrative intervention patterns.

This is demonstrated in Snellen's survey of new forms of governance. He distinguishes between processes of communal image-creation on the one hand, and parametrisation, proceduralisation, and structuring of steering and decision-making arrangements on the other (Snellen, 1987: 21–25).

Image-creation as a mode of steering is not essentially policy-making. It is a process of organising and stimulating, of encouraging a 'rich' exchange of ideas between interested and involved actors. Such steering starts out from the notions of 'not-knowing' or 'unable-to-know'. Relationships and developments in a policy domain are so unpredictable, complex and changeable that no single actor – in this case, government – is in a position to produce meaningful images. Also, because of professionalism and technical complexity, the policy domain has become so impenetrable that meaningful images can only be produced within the changing and fragmented structures relevant to the domain itself.

Steering by parameters, procedures and structures is a form of abstract governance, of meta-steering, in which the administration chooses the financial or decision-making infrastructure of a domain as the point of application. This infrastructural steering is also indifferent to outcomes. However, it should be noted that with parameter-steering – especially if it involves defining achievements or achievement-indicators – certain fundamental preferences can still apply.

The result of these new forms of steering is that, although intended to be a response to fragmentation, they in turn encourage further fragmentation. After all, they link up with the variety of image-creation in different policy domains and simultaneously try to define or acknowledge very specific financial and decision-making infrastructures. The process of postmodernisation thereby gains more momentum, and the administration fragments even more.

Governance and Cyberspace

Information and communication technologies are a product of modernisation. They embody the culture of functional rationality and are employed in strategies of rationalisation, discipline, and control. In that sense, ICT, when applied to bureaucracy, strengthens the image that the political masters are in control. On the other hand, there is an immanent tendency in ICT to sideline the political masters. Informatisation is always a process of technocratisation.

But ICT, certainly at the present time, is undergoing a number of developments which are reducing its bureaucratic characteristics (see Frissen, 1989). These developments, described in Chapter 3, are: increased quantitative and qualitative capacity; the potential for linking databases; network expansion; and developments in virtual reality. They have fundamental consequences for public administration. The increased capacity of systems and applications makes it technically possible to opt for the small-scale and reduces the need for centralisation and specialisation. Small, decentralised units can now be equipped extremely powerfully. Furthermore, forms of organisational intelligence can be linked directly to policy implementation. Hierarchical feedback is not only unnecessary but often counterproductive and inefficient. Intelligent fragmentation is now feasible and the result is subdivisional autonomy within organisations. The monocentric organisation is becoming ever more obsolete. Data-coupling and networking stimulate a pattern of horizontalisation in and between organisations in the administration. Such horizontalisation is in part the result of deliberate restructuring and specific interventions; an effect, for instance, of meta-steering. But even more is it the result of autonomous technological development on the one hand, and the spontaneous outcome of actions by technologically intelligent users and organisations. This explains the impassioned pleas for more secure information infrastructures in public administration. However, contingencies keep finding new detours and inside tracks. It is more sensible just to accept the situation.

Data-linking, networking and virtual reality have left the notion of 'territory' behind, at least as the primary rationale for the design and activities of public administration. Time, distance and space are becoming relative concepts through which the functioning of organisations is becoming de-territorialised. The Dutch Inland Revenue service could easily offer its services elsewhere in the world without any significant physical displacement. In an age of ICT, territorial redistribution in public administration, by creating 'urban provinces' for instance, is an anachronism resting on the classical notion of monocentric bureaucratic order and political primacy.

The result is functional differentiation and the virtualisation of organisation structures. Organisational boundaries become blurred and their physical location becomes irrelevant. Integrated steering, underpinned by a vertically integrated politico-administrative system, is little more than a dream unless it is accepted that any coherence will be a temporary and accidental outcome which could only be observed retrospectively.

Virtualisation also highlights the proliferation of images and styles produced by ICT. Their transience and changeability mean that social relations acquire a certain anarchistic character. Control, like predictability, has become an illusion. This requires a fundamental reorientation of public administration in cyberspace.

At the same time, the realities of cyberspace are becoming increasingly a reality of policy-making. That is only to be expected in terms of technological policy and the various programmes needed to stimulate the digital highways as a new infrastructure. But the new infrastructure is becoming more and more a precondition for steering and its point of application. After all, since an important part of public administration is directed towards information-provision and communication, the technologies which have been developed, or are in development, in that field will have to constitute its core. It seems to me self-evident that the characteristics of technoculture described above will influence the codes of meaning in and through the administration. Here too one can speak of a process of postmodernisation in public administration: decentring, proliferation of images, fragmentation, ephemerality and changeability, de-territorialisation.

8.4 POLITICS: PRIMACY OR THE END?

In political-administrative theory, the political system stands at the crossroads of administrative and technological development: it provides meaning, creates images, gives direction, weighs up and makes choices. It has the role of a centre which, although not dominant in every respect in a parliamentary democracy, nevertheless ultimately represents and directs society. However, it is at these crossroads that administrative and social developments are making new connections which threaten the central position of the political system. As has been observed earlier, the ongoing fragmentation can not tolerate the idea of a centre, let alone a single centre. That applies to meta-narrative (Lyotard, 1987a) as well as monocentric organisation (Crook et al., 1992; Clegg, 1990). The political-administrative system is the clearest expression of both those phenomena.

This perception, expressed with varying degrees of hesitation, is encountered in a number of recent publications: Bovens, Derksen, Witteveen, Becker & Kalma (1995), Depla and Monasch (1994), Guéhenno (1994), Van Gunsteren and Andeweg (1994), Huyse (1994), Mulgan (1994), Tjeenk Willink et al. (1994), Tops (1995). The political-administrative response can be seen in recent political pleas in favour of restoring or acknowledging the primacy of politics. But a system which has to redefine its *raison d'être* is without doubt a system in crisis.

Fragmentation and the Centre

An inherent effect of the modernising process is to challenge the position of political institutions. Modernisation means, after all, the advance of functional rationality and a relative loss of substantive rationality. In the political-administrative system, it is politics which pre-eminently represents the normative component of substantive rationality. So, in a sense, the disappearance, diminution or greater abstraction of politics is modernist logic. Again we are back to bureaucratisation and technocratisation.

This immanent tendency is primarily a shift of power: between politics and bureaucracy, between politicians and bureaucrats, between those entrusted with authority and technocrats. This power shift often takes on the character of a change of presentation, of image, of ethos. So administration then becomes 'to instrumentalise' or 'to economise' (Foqué in Tjeenk Willink et al., 1994: 89–92); politicians and politics are bureaucratised; politics becomes formalised and pragmatic (In 't Veld in Tjeenk Willink et al., 1994: 103–105). The positions and players change less than the codes of meaning, the discourse, the style. Democracy becomes a war of images, a fact which many find totally unacceptable. However, this shift of power does not end here. Guéhenno (1994) sees the demise of democracy primarily in the pluralisation of power centres which results from the increasingly networked character of modern society. This network character grows exponentially through informatisation. Politics, in the sense of public decision-making or decision-making which is relevant to the public, shifts or is displaced. In different words and with different degrees of approbation this has been observed by Bovens *et al.* (1995), Depla and Monasch (1994), Huyse (1994) and Tops (1995). Bovens *et al.* list six developments which cause this displacement: internationalisation, regionalisation, bureaucratisation, technologisation, individualisation and judicialisation (Bovens *et al.*, 1995: 13-20). They claim that:

> The displacement of politics is in many instances a displacement of social power and not, or not yet, of democratic control. Many of the new political arenas are closed clubs. The danger is that the achievements of democracy, such as openness and responsibility, will be sacrificed on the altars of internationalisation, economic rationality and technocratic effectiveness. (Bovens *et al.*, 1995: 21)

But these shifts and displacements might still just be seen as a pattern of competition and changes of position. More fundamental are the joint effects of administrative and technological developments, especially when combined with social developments.

The process of virtualisation reduces the significance of territory as a normative and functional basis for the political system. Virtualisation arises through the horizontalising and autonomising aspects of administrative development and is powerfully stimulated by technological development. Politics, as the decision-making centre of a territorial unit, ends up in a vacuum.

Virtualisation leads, moreover, to organisational fragmentation. The scale of organisations declines through fragmentation, division, privatisation and contracting-out. Vertical integration also declines, making way to contractual links and network configurations. Fluidity and change increase through the blurring of organisational boundaries and constantly changing arrangements. (See also In 't Veld in Tjeenk Willink *et al.*, 1994: 97ff.)

The significance of this is that politics no longer automatically has at its disposal the loyal instruments and servants of power. Where bureaucracy is fragmented, the exercise of political power evaporates and is replaced by uncontrolled power, contingencies or competing political forms.

Finally, virtualisation also reduces the relative importance of meta-narratives. Disbelief in meta-narratives is in any case more widespread, but it has fundamental significance for politics. Politics is a grand narrative at two levels: first, there is the grand narrative of politics as the centre of public decision-making; secondly, politics is an ideological struggle between competing grand narratives. The disappearance of political ideologies means a loss of function and normative erosion for political institutions, politicians and political parties. Shifts and displacement in politics undermine the entire political system more powerfully because the place of power, however empty it may be, can no longer be located.

Virtualisation, in other words, is a process of fragmentation which causes the centre to disappear both at the level of the individual organisation as well as at the level of society. Consequently, the central position of politics for both bureaucracy and society is eroded. This dual erosion is not compensated for by the appearance of a new centre. Usually many centres emerge, but even more likely is that the significance of the centre as such becomes empty. Again we are observing a process of postmodernisation: fragmentation, ephemerality, plurality of images and a disbelief in grand narratives.

Trivial, Arbitrary, Harmful

These developments are problematic, even tragic, for the political system. An awareness of this comes through in the debates and self-reflection described in Chapter 5. But no clear conclusions have been drawn. Instead people have continued down well-worn paths and consoled themselves by playing down the gulf between politics and society: a gulf can not be demonstrated quantitatively, or it has always been there, or it is universal (Van Gunsteren & Andeweg, 1994).

Nevertheless, it remains tragic and requires further explanation because tragedy implies the unresolvable and reflects a situation which transcends the purifying powers of actors and systems.

I would describe the position of the political system within a process of postmodernisation in three ways: trivial, arbitrary or harmful. It has become trivialised by two developments. Firstly, it is the result of modernisation's inherent tendency to cultural generalisation (Zijderveld, 1983). Through functional differentiation, the values which are central to substantive rationality become abstract and relatively meaningless. We can see it in the clichés of so-called 'mission statements' released by organisations and which, *a fortiori*, is also true of the mission statements which politics would have to formulate for society. So to advocate a restoration of the argumentative qualities of politics and democracy (Tjeenk Willink *et al.*, 1994) is somewhat naïve. It fails to understand that the arguments are pluralised and fragmented and would inevitably lead to a narrow localising of content. And while Bovens *et al.* rightly observe a displacement of politics, their conclusion that 'this in no way diminishes the importance of ideas in giving political direction' is largely wishful thinking (Bovens *et al.*, 1995: 57).

Secondly, trivialisation occurs through the disappearance of grand narratives and a narrowing of the differences between political movements: the Dutch one-party state, as Oerlemans described it. This refers to the position of different actors such as political parties, but even more to that of the political system as a whole since its fundamental legitimisation by ideological choice, sanctioned by the majority, is absent. What is sanctioned is trivial.

The arbitrary position of politics follows from this. As the normative basis of politics is levelled down and trivialised, its monopoly position in public decision-forming becomes arbitrary. From what can its exclusive position be derived? What normative objections can now be raised against the displacement of politics? It does not help to justify the centrality of politics tautologically, as Wöltgens (1992) does, by pointing to the permanence of political or public issues and subjects within society. Furthermore, politics also becomes arbitrary through the complexity and contingency of social developments.

> Für das politische system als gesellschaftlichen Ort der Entscheidung über verbindliche Regeln ist die Denkbarkeit von Kontingenz ein Danaiden-Faß. Zwar verdankt sie dieser neuen Form des Denkens über Regeln ihre Existenz als sekuläre Ordnungsform, aber nun gerät *jede* realisierte Ordnungsform unter den Druck möglicher Alternativen. (Willke, 1992: 36)

Any choice between contingencies is arbitrary. And Willke too points out its tragic character, because the right of politics to exist is derived from the making of choices within a contingent reality. And whenever politics resists its trivial and arbitrary position, it immediately does harm. For it then has to emphasise its ideological nature or its exclusive position. The ideological aspect implies the telling of a grand narrative while exclusivity means restoring the idea of a centre. The grand narrative which it tells inevitably tends to become bureaucratic or technocratic whenever it is concretised in terms of policy or regulations. And it has to be concretised because generalised ideological arguments in postmodern reality are mere abstractions. And it is necessarily bureaucratic and technocratic because they are the only arguments which lead to adequate functional differentiation and control. We can see immediately the management thinking of the 'large-scale', as for instance in recent plans to restructure health and education in the Netherlands.

The damage which this causes is to threaten the destruction of variety, which in a fragmented society is countered by evasion, calculation and further fragmentation. At the same time, the exclusive position claimed by politics leads to growing conflict and competition which results in declining legitimacy and further fragmentation. The vicious circle is indeed tragic. Ankersmit, a writer normally sympathetic to postmodernism, seems to have overlooked this when he calls for the restoration of 'a strong state' (Ankersmit, 1994a).

Political Primacy

The rehabilitation of politics is advocated on two levels. On the one hand, at the level of decision-making – arguments for political primacy – and on the other, at the normative level – arguments for a return to political discussion about values and norms. These arguments can be heard in political, administrative and academic circles.

The restoration of political primacy was an important credo of the 'purple cabinet' of 1994. The viscosity of decision-making was assumed to have been caused by the congealed decision-making structures of corporatism, the opaque interweaving of government organisations, society's mid-field and private sector, the system of checks and balances and expanded legal protection. The position of politics as the decision-centre was also thought to have been undermined by the fragmentation of public administration through, for example, privatisation and independent agencies.

So if politics was to win back its primacy, it would have to be achieved by a clearer demarcation of public and private responsibilities, by disentangling corporatist arrangements, by reorganising procedural and structural guarantees, and by uniform and harmonised organisational structures.

In the administration, too, we can see attempts to restore the primacy of politics. I summarised this briefly in Chapter 5: discussions about core-tasks; the notion of outline governance; the division between policy development and policy implementation in devolving independence; demands for greater efficacy. Innumerable attempts at administrative reorganisation – tellingly illustrated by restructuring plans and the forming of urban provinces – can be seen as a re-ordering of a fragmented, administrative reality in order to make the exercise of political primacy effective. Incidentally, the terminology still remains technocratic: administrative integration and scaling-up, control and responsibility.

A striking feature of the academic debate is a tendency to the normative. The empirical diagnosis that political primacy is being undermined by displacement, shifting and fragmentation has met with little criticism, apart perhaps from Hoogerwerf (1995). The rest accept the diagnosis or make it themselves, e.g. Bovens *et al.* (1995), Hupe (1995), Tjeenk Willink *et al.* (1994). But this empirical fragmentation is countered normatively. Bovens *et al.* and Tjeenk Willink *et al.* advocate strengthening the normative and argumentative dimension of politics. Communitarianism and republicanism, too, can be seen as attempts to find a new place for morality in politics and thereby to legitimise anew its position of primacy. Re-ideologising politics should, after all, make it less trivial and arbitrary and give substance to its role as the ultimate centre of public decision-making.

But even allowing for the fact that the empirical is a mental construct, we should not ignore it. If we recognise

> (...) that if there is a pattern of change it is in the fragmentation of old cultures and the proliferation of new values, attitudes and attendant behaviour, lifestyles and political movements in their place (...), that the emerging character of

contemporary political culture is pluralistic, anarchist, disorganized, rhetorical, stylized, ironic and abstruse (Gibbins, 1989: 23),

then resorting to the normative is a despairing attempt to rescue modernity. In that modernity, politics is the secular body which makes decisions about the future development of society. It does so on the basis of a normative, ideological narrative and is in the exclusive position of being able to formulate the general interest. But this aspiration is modernist because it remains stubbornly decisionistic and because it refuses, and in fact is unable, to conceive of any development towards decentring.

It is, moreover, in a sense tautological. Political questions and subjects naturally continue to exist in postmodern society, at least if we accept that politics is a specific vocabulary and a word-game. Politics is a historical and cultural construct. But in no way does it mean that existing political institutions have a monopoly of these issues and subjects: this would virtually imply a natural exclusiveness which can only arise from an essentialist position. And essentialist positions are always tautological because ultimately they are self-referential. Neither is it possible to maintain that politics remains one and the same at all times, and that it will be based upon morality and normativity. Powerful alternatives, however, are to be found in intelligence, aesthetics and pragmatism.

8.5 POLITICS, GOVERNANCE AND TECHNOLOGY: FRAGMENTATION AND CONNECTIONS

The developments taking place in politics, administration and technology are connected by fragmentation. This fragmentation is not unique to these domains, as we saw in the sketch of the social décor (Chapter 6). Theoretically, the fragmentation is expressed in the notion of postmodernisation, a process which shows elements of continuity and discontinuity with modernisation. The dual importance of this continuity and discontinuity is symbolised by my use of the term 'fragmentation' as an alternative to 'de-differentiation' which some writers prefer (Crook *et al.*, 1992; Lash, 1990). Fragmentation is the consequence of an ongoing process of differentiation within modernisation which is ultimately radicalised into hyper-differentiation. This hyper-differentiation puts an end to the linear process of differentiation emanating from a centre. That centre may be the top of a pyramidal organisation, the subjectivity of an engaged actor, the sovereignty of a political decision centre, the idea of enlightenment, progress and reason, or the core of a grand narrative.

Fragmentation expresses the disappearance of a centre and linearity on the one hand, and the emergence of new connections on the other. Those connections are kaleidoscopic fragments drawn from various domains. Their emergence is therefore largely contingent and their form is more important than their content.

Changing Administration and Problematic Politics

We can speak of a process of postmodernisation in public administration. The fragmentation of public administration is, on the one hand, a more or less contingent result of bureaucratic differentiation and transformation, and on the other, a form of intelligent design for achieving some correspondence with social fragmentation. We can see that in the developments described earlier, such as localisation, functionalisation and autonomisation, which give public administration (in the long term if not already) the form of an archipelago, policy implementation is becoming more intelligent and strategic, and self-steering is no longer a problem but a basic principle and objective.

The increasingly networked character of public administration is leading to the growing importance of horizontal connections. This again is partly a design choice but also a contingent result of the openness and non-hierarchical structure of networks. Furthermore, aesthetic aspects are becoming more important in assessing the effectiveness of policy co-production in networks. Achieving consensus and acceptance and avoiding conflict in an elegant fashion are more important than the actual content of the result. This again means that effective administration has become a question of style and design.

This postmodernisation of administration and the social context in which it operates is problematic for politics. Postmodernisation problematises politics in a way which affects both its position and its content.

The position of politics is affected by fragmentation in two ways. In the first place, social fragmentation makes an exclusive centre for public decision-making impossible and undermines its legitimacy. In the second place, the fragmentation of public administration leads to a decentring of the political-administrative system. As a result, politics is having to manage without its classical set of control instruments. If it tries nevertheless to maintain its central position, it gets swallowed up in bureaucratic and technocratic discourse.

Postmodernisation makes the content of politics problematic, which is to say trivial and arbitrary, because the grand narratives of ideology no longer enjoy a binding and legitimating effect. That is not to say that ideologies or value patterns disappear, but that they have to compete with innumerable fragmented lifestyles and other codes of meaning. Decentring and fragmentation of position and ideas occur. Politics, in its ideal-typical form, seeks ideological consistency and permanence and can not cope with ephemerality and contingency. Too much plurality inherently reduces the importance of overarching ideology as well as the possibility of centralised order, decision-making and direction.

The contribution of technological developments to the postmodernisation of administration and politics is obvious, but ambiguous. There is always a temptation to think that technology can produce perfect control. Big Brother is not just a dream; in many domains it is a reality and often resulting from the noblest intentions. And yet important technological developments are pointing in an entirely different direction: increased capacity, networking, linking, virtual reality. It is that direction which has most in common with postmodernisation.

The wide application of ICT in public administration means that some aspects of the technoculture which characterises our society also influence the administration. Decentring the subject diminishes the role of the individual decision maker. This can be seen, for example, in the imposition of fines for traffic offences. (See also Snellen, 1993; Zouridis & Snellen, 1994.) Other developments are taking place in the areas of decision-support, scenarios and policy exercises. Decentring the subject also means that organisations become decentred. If we see them as a collection of 'man–machine tandems' (Lenk, 1994: 318), any centralised control is an illusion. Government organisations are thus becoming more fragmented and less steerable.

This fragmentation is further encouraged by the unbridled image-production of the technoculture. Networks, integration and linking, certainly in connection with virtual reality, expand not only the number of competing images but also their reality-value. The situation definitions generated by ICT acquire an autonomous significance in the processes of agenda- and policy-creation (Meyer, 1994). That means not only a bias to computer control but also to pluralisation of bias.

Networks and links lead to a horizontalisation of organisational relations. The Internet symbolises this. So ICT accentuates any horizontalising tendencies brought about by restructuring and reorientation in public administration. They in turn contribute to decentring and fragmentation, especially since many forms of horizontalisation have a contingent character.

These developments can also be described as a pattern of virtualisation. As the compression of time and space reduces their importance, the design of policy processes and organisational structure is altered. Functional differentiation and integration, specialisation and de-specialisation, separation and connection of policy development and implementation are all, and simultaneously, possible. Through ICT, the co-ordination of functions and activities no longer has to be bureaucratic and hierarchical.

Scale is less important as a criterion for the design of the administration. Instead there is 'glocalisation' whereby large-scale activities can be performed on a small scale and vice-versa (Virilio, 1995). The importance of territory declines. In this way, virtualisation enhances the timeless experience of immediacy: everything is available and transparent in the here and now. That is an aspect of the staccato character of the technoculture. It strengthens fragmentation, but at the same time stimulates the seductiveness of transparency. Immediacy appears to allow the boundaries of knowledge to be transcended and to place total control within arm's reach. At least that is the ambition which it feeds.

The transformation of public administration, which ICT has accelerated, has created further problems for politics. The decentring of subject and organisation has two consequences. In the first place, it weakens the instrumental characteristics which, in principle, are expected of an ideal-typical bureaucracy. In the second place, it affects the position of politics as the public centre of decision-making. The image of the sovereign – the political representation of society's subjectivity – crumbles. Not only does technological 'embeddedness' increase (Van den Hoven, 1994: 359), but the number of centres – subjectivities – are multiplied. Any talk of a single centre for public decision-making should be treated with irony.

The technoculture's proliferation of images is competing with the claim of politics to a monopoly of public image-forming. Not only is public image-forming being fragmented, but image production is so ephemeral and changeable that the centre of political consideration and selection can hardly be found. The coagulated images produced by political decision-making become immediately obsolete, and thus trivial and arbitrary. Any grand narrative has become impossible.

Horizontalisation also makes the position of politics problematic. The present anarchy on the Internet has seen the emergence of virtual politics and virtual political communities (Rheingold, 1994). Closely related is cyberpunk which, according to Burrows (1995), is a new political theory but with no desire to formulate a utopian vision.

But over and against cyberpunk, there is the potential to realise Big Brother ambitions. Intensive control and universal transparency continue to thrive as ambitions. And to repeat, they can result from the noblest of intentions.

Despite such ambiguities, whose outcome is still uncertain, I believe that the most likely result is the virtualisation of the political system. The characteristics of technological development virtually exclude the emergence of dominant meta-narratives and make the success of universal and unlimited control highly doubtful.

Much more do we see a pattern of de-territorialisation through processes of time–space compression. A delimited territory and linear chronology are no longer needed as the basis for the political system. Furthermore, each temporary creation of that basis is made obsolete by new technological and social developments. This means that connections become contingent and will only rarely obey the logic of the political system.

Immediacy of experience and activity is increasing dramatically. Peters (1992) cites a financial specialist who argues that, in global terms, ten minutes is a long time. The traditional mediating role of politics – in image-creation, debate and decision-making – is set under pressure in spite of the fact that immediacy also feeds the ambition to control and dominate.

Digital Ambiguities 3: Public Administration and Political Theory

So postmodernisation alters public administration and makes politics problematic. There is a striking resemblance between administrative developments, summarised earlier as change, experiment and innovation, and certain specific characteristics of technological development. Both developments lead to a process of virtualisation: horizontalising, autonomising, de-territorialising and decentring are but some of these aspects, which in turn correspond to the dimensions of postmodernisation described in the theoretical intermezzo.

The significance of postmodernisation, however, is problematic for the *position* of politics. The notions of meta-narrative, a public decision-centre and territorial unity associated with modernist politics become obsolescent in the process of postmodernisation. Herein lies the tragedy of politics: its natural life is outliving its institutional life.

And there are many ambiguities. The political lifestyle of cyberpunk continues to oppose the persistence of Big Brother ambitions; the process of virtualisation is countered rhetorically by the advocacy of political primacy; and as an alternative to fragmentation there is the re-ideologisation and moralisation of politics.

Public administration and political theory reveal these ambiguities to the same degree. After all, both are interwoven with the modernist tradition: public administration as the science and technology of control and problem-solving; political theory as a provider of grand narratives and justifications. At the same time, they both point forward to postmodernisation: public administration primarily on the grounds of empirical research which has revealed deficiencies in the modernist urge to control which bureaucracy embodies; political theory on the grounds of the problematic legitimacy of the state, arising both from its ineffectiveness and from philosophical objections to its moral and ideological foundations.

The answers are self-evidently ambiguous. The recognition of ineffectiveness leads to the formulation of a more sophisticated conception of steering and policy. However, public administration repeatedly runs up against the vicious circle of the monocentric bureaucracy under the primacy of politics. The recognition of problematic legitimacy leads to repeated attempts to redefine public morality. Political theory, thus, constantly runs up against the fragmentation of postmodernisation.

So the question which was phrased in general terms in Chapter 7 must now be focused more narrowly on public administration and political theory. If it is possible to speak of postmodernisation in society and the worlds of politics, administration and technology, should there then not be a postmodern theory of public administration and postmodern political theory? The question is all the more urgent because public administration, being an object-oriented discipline, ought to be able to process the significance of postmodernisation both descriptively and prescriptively, and because political theory as a provider of justifications based on solid arguments ought to take a stand in the face of postmodern vocabulary. In the final chapters of this book I shall attempt to supply an answer to these questions. I shall be using the terminology of postmodernism in my discussion of the theory and practice of administration and politics.

9 Fragmenting and Connecting Governance

A postmodern narrative about the postmodernisation of public administration: does this adequately describe a postmodern science of public administration? And can it do justice to the picture of fragmentation and interconnections which previous chapters have sketched? It remains to be seen. This chapter is an attempt to tell that story. It will have to be judged on its own merits since I know of no universal criterion by which it can be judged or which, within the story itself, would not be rejected as impossible or dangerous. Anyone who believes that fantasy in academic writing is impermissible, or should be subordinate to methodological discipline, uses a different vocabulary from myself.

It is inevitable that this narrative will often treat the theory and practice of administration as one. Any science is deeply interwoven with its object of study. And however far we may try to distance the academic discipline of public administration from its object – elsewhere I have used the phrase 'cool science' (Frissen, 1991) – its significance will always depend on its relevance to the real world of public administration and administrators. If an academic produces a text on public administration – it is after all his main activity – he thereby adds to the textuality of public administration (see also Van Twist, 1995: 79ff.). The significance of that text lies in its relationship with other administrative and theoretical texts. Its significance will certainly depend to some extent on its 'performativity' (Lyotard, 1987a). Texts are 'true' if they impact on the dominant discourse and succeed in transcending its boundaries. Creating images of public administration (Idenburg & Van der Loo, 1994) is an important task for administrative theory. But then fragmentation must be acknowledged:

> Instead of the 'terrorism' of integrating truths and images, people in the postmodern era now have to deal with the stimulus of many divergent opinions and interpretations of reality. While this has made life more difficult, it has also made it infinitely richer. People are less dependent than before on integrating images imposed, as it were, from above. They are now encouraged to pick up numbers of small image fragments and edit them into a film (Idenburg & Van der Loo, 1994: 27).

Editing a film is to create something new from a number of connections. And that is precisely what happens within the fragmentation of postmodernisation. Connections come about haphazardly, and the combination of fragmentation and connections forms the central theme of this narrative about the theory and practice of postmodern public administration.

9.1 FRAGMENTATION AND CONTINGENCY

Fragmentation is often perceived negatively as a problem. Cultural sociologists such as Zijderveld (*e.g.* 1991) deplore the staccato character of the culture which it gives rise to. Even a specialist in public administration like Hoogerwerf stresses the negative side of fragmentation and its status in postmodern thinking (Hoogerwerf, 1995: 151–152). Naturally, those who are attracted by variety, observe it all around them and do not consider it *a priori* to be problematic. However, there are other arguments in favour of fragmentation than mere personal preference.

Intelligent Ephemera

Every transaction is tied to time and place and that is why Lyotard talks of 'local determinisms' (1987a: 27). Every administrative action is therefore specific and can not be explained by generalised theories (Abma, 1994). The variety and the local determinants of administrative actions are further intensified by the process of time–space compression, as we saw earlier. Through this process, the traditional constraints of time and space, such as temporal linearity and the need to bridge physical distance, become relatively less important. There is now a simultaneous process of both compression and acceleration. The plurality of reality is present in all its immediacy and at the same time changes extremely rapidly. The ephemerality of social conditions reflects this, and it is also true of activities, processes, image creation and problem definitions. The consequences for patterns of administrative intervention are threefold: in the structuring of public administrative organisations; in the design of policy processes; and in the production of images and the organisation of communications for them.

The ephemerality of social developments demands greater flexibility in structuring administrative organisations. Organisational structures have to become more intricate and more closely targeted on the specificity and changeability of social domains. The process is comparable to that of flexible specialisation in the business world. In general, this demands greater responsiveness and adaptability from administrative organisations, which generally means splitting them up and reducing their size. Where this does not happen in a physical sense, it nevertheless occurs through more internal independence and autonomy. These developments are well documented in the research literature. Bekke, Hakvoort and de Heer (1994) discuss departments; and there is the work of Van Twist and In 't Veld (1994) and Depla (1995) on local administration. The crucial thing is that this restructuring is a permanent process which moreover receives a further impulse from ICT because, thanks to electronic networks, relational databases and communal information architectures, organisation patterns are now better able to follow the network structure of social domains.

The main function of organisational restructuring is to facilitate the design of policy processes. Those policy processes have to respect the fragmentation of social domains. That means that they must be so designed that they can process a wide variety of actors, shifting coalitions and conflicts, and changes in the ground rules. This requires a high degree of openness and procedural flexibility. Interactive processes are crucial, so focusing on results often has to be played down, especially when the common interests of the participants have to be kept in mind. The interaction as such rather than as a concrete, formalised result will often be defined as the policy outcome. Also essential is a wide range of policy instruments, with a preference for 'second generation steering instruments' (De Bruijn & Ten Heuvelhof, 1991). New forms of steering, new concepts of administration, are sometimes the inspiration for the redesign of policy processes, though more often they are a result of it (Bekkers, 1993; Bekkers & Van de Donk, 1989; Frissen, 1990; Snellen, 1987). Steering can only be successful if it can process all this complexity and dynamism, and there will usually be a large element of self-steering. (See Kickert, 1991.)

In order to keep policy processes and steering arrangements sufficiently open, it is necessary not to resist, either by selection or reduction, the proliferation of images generated by social fragmentation. This involves acknowledging the plurality of images not just in a technological sense, for instance through scenarios, policy exercises (Geurts, 1993), communal image-creation (Snellen, 1987), but also learning to manage them administratively as a desirable outcome of the policy process. It demands an active acceptance of the ephemeral: it is a characteristic of social developments, and should be a basic principle for policy-making and steering. Accepting the ephemeral is unavoidable and therefore technocratically intelligent. Ephemerality is inextricably bound up with plurality; and for that reason accepting it is also aesthetically intelligent. Only by such acceptance does it become possible to seek new connections.

Necessary Relativism

Recognising and acknowledging social fragmentation without striving to reduce variety, implies that the administration should adopt a relativist position. However, relativism is often, though wrongly, associated with indifference. There are three meanings of relativism which are relevant to administration.

The first meaning of relativism is the recognition of the plurality of codes of meaning and in particular of their dynamic. Rorty talks of vocabularies which are not reducible to each other. It is thus administratively intelligent to design organisations, steering processes and image-creation so that this plurality can be seen and respected. That will mean that public administration expends less energy on creating dominant and uniform images as the most important goal of policy-making and steering. In this I differ fundamentally from Van Twist and In 't Veld when they argue that core departments should focus on 'improving the coherence of policy at the level of the state' and 'raising the quality of the policies and the steering' of implementing organisations (Van Twist & In 't Veld, 1995: 206–207). The goal of administration should be to seek out variety and give due weight to competing policy-images; why should one always have to make a choice? Of course, that will mean further fragmentation in public administration, but from a social perspective it would be more effective and from a normative standpoint of checks and balances it would be more intelligent. In this way too, competition and rivalry can be comprehended.

The second aspect of relativism is the rejection of the concept of an integrating value or criterion. As Rorty argues (1991a: 202), this rejection implies that it is unacceptable either to search for some ultimate truth or dominant value, or to treat all 'vocabularies' as essentially the same. Relativism is far from an obligation to do nothing; it does not mean apathy. Actors can negotiate, reach agreement and accept outcomes. But there is no need for any moral legitimisation by which one outcome is considered to be right or just or is accorded some measure of moral superiority. Administratively this means that policy processes are designed so that negotiation becomes acceptable, that agreement is sought and that the significance of the outcome is seen to lie precisely in the process of negotiation and agreement. Relativism, when understood in this way, leads to a pragmatic administrative vocabulary. Or in other words, muddling through. Every regulation is relative and negotiable because regulation is always the result of negotiation.

The third aspect of relativism lies in its hidden potential for creating relationships and making connections. Acknowledging that social reality is fragmented also involves an awareness that every fragment is a partly created and partly contingent set of connections. To be sure, the emergence of new connections is uncertain ('indeterminacy' is an important postmodern concept : Cooper & Burrell, 1988: 98 ff.), but it can also be the input for administrative actions. A relativist perspective means that administrative actors are indifferent to the content of the connections – they will always remain uncontrollable and accidental unless 'totalising' pretensions are brought into play. Administrative actors must focus on setting up processes and structuring organisations in such a way that while they recognise and respect the indeterminate outcome of social fragmentation, they will also stimulate the rise of new connections. The emergence of a connection is the 'proof' of effective steering and policy-making. Processes become more important than their outcomes.

Relativism, then, is not only inevitable, it is also necessary. In the three aspects which have just been highlighted, relativism in an administrative sense leads to an acknowledgement of plurality. Administration is then focused primarily on maintaining plurality and the conditions needed for plurality (see also Van Gunsteren, 1993) just because within it are contained the possibilities of creating connections. To return again to Rorty, avoiding cruelty, conflict which is destructive of variety, and senseless violence, is an important administrative goal. That is ambitious enough. Meanwhile relativism (irreducible vocabularies, no hierarchical criteria, the possibility of connections) remains the basic principle.

Unavoidable Contingency

Social fragmentation and especially the dynamic emergence of fragments as intended or accidental connections confront the administration with the inevitability of contingency.

> Der langwierige Prozeß der Verdrängung der Religion durch die Politik in Zuge der (...) Säkularisierung gab anderen Differenzen oder Antonymen eine Chance - bis zu dem heute wieder beklagten Punkt der Relativität und Relativierung *jeder* Differenz. (Willke, 1992: 36)

Willke argues that the administration has to recognise contingency in order to gain some level of control over it. For him the danger of contingency is that it can be self-destructive. Hence the need for control. But such control needs to be based fundamentally in keeping open the possibilities for contingency. It means that public administration should not start out from a preference for consensus and the primacy of politics which that implies, but should give the primacy to dissent. Only this can guarantee further differentiation, and from the theoretical position adopted by Willke, it is a necessity.

> Die Rede vom primat des Dissensus soll sagen, daß das 'Bewegungsgesetz' komplexer, eigen dynamischer Sozialsysteme zu sehen ist im *regelgeleiteten Prozessieren von Differenzen und Dissensen.* (Willke, 1992: 49)

The point of departure is that everything changes except change itself and that administrative systems must control contingency by regarding it as both a basic principle and a desired result. In this I can agree with Willke but I am less happy with his theoretical approach which has a strong functional orientation. Although thinking in terms of functions is intended primarily as a heuristic device, it hides an ontological connotation. Functions and functionalism all too soon acquire an essentialist interpretation. How far that can co-exist with the notion of contingency is questionable.

For that reason, I prefer Rorty's conception of contingency which is less functionalist and more historical. Furthermore, Rorty does not suggest that the aim is to get a grasp of contingency: it is enough merely to be aware that language, consciousness and community are contingent. The pursuit of connections and involvement will then always be accompanied by an awareness of their accidental nature (Rorty, 1989). In administration it means that, analogous to the 'law of the required variety' (see Kickert, 1991: 21), contingency should be a hallmark of steering. Any desire to co-ordinate and integrate all administrative activity thus becomes disruptive. It is better to leave the contingent 'balance' of various administrative actions as contingent, and to abandon the idea of total control. Once again, the Internet shows how contingencies can lead to coherent results. But the core of coherence on the Internet is precisely this contingency. Regulation and organisation to combat these contingencies destroy their coherence because they destroy variety. Regulating and organising hybrid patterns (the Internet, for instance, but also social domains in general) can only be done meaningfully by the patterns themselves. And any sense they acquire will lie in the contingent nature of the regulation and organisation. This makes autonomy and self-steering important principles of postmodern administration and postmodern administrative theory.

9.2 AUTONOMY AND HORIZONTALISATION

Ephemerality, relativism and contingency are a result of fragmentation and imply the possibility of connections. Public administration is itself also fragmented because it seeks to key into social processes intelligently. Relativism, in the three aspects described earlier, becomes necessary because thereby fragmentation will no longer be perceived as problematic and it will become possible to make connections. The establishment of these connections, however, is usually contingent even when they are sought. This contingency is often preferable to conscious design, because the possibility of a more intelligent or more elegant outcome is left open.

The consequence of such an administrative orientation is that actors, processes and domains acquire an autonomous character and the connections which arise are generally horizontal. That is to say, they are non-hierarchical in either an organisational or a political sense. At the same time, the autonomy of actors, processes and domains and the horizontalisation of relationships can be either the causes or the conditions for that administrative orientation.

Necessity and Trust

One of the core perceptions of autopoietic social system theory is that actors, processes and domains in our modern, functional, highly differentiated society are autonomous. For public administration and politics, the work of Luhmann, Teubner and Willke is especially relevant. In the Netherlands, there is the volume of In 't Veld, Schaap, Termeer and Van Twist (1991). One does not have to accept the entire system-theory of autopoiesis to be able to take pragmatic advantage of its insights. Even its opaque and irritating jargon need not prevent us from picking out some interesting ideas. Fortunately too there is Van Twist who as always writes lucidly:

> Autopoietic systems are not open systems which transform an externally defined input into an output, thereby adapting continuously to changes in the environment. Autopoietic systems are self-referentially closed; they reproduce themselves in a closed system of reproductive relationships. In that respect, autopoietic systems are autonomous. The environment is unable to determine a self-referential closed system; it can only disturb the self-reproduction of such a system and supply peripheral conditions. Cause–effect relationships between the system and its environment always have to pass through a barrier of internal, self-referential reconstruction. The relationship between system and environment is therefore not one of causation but of restraints. (Van Twist, 1995: 193)

Van Twist argues rightly that autopoiesis is concerned with the autonomy of communication. He links this notion to postmodernism which is non-representational. In that sense, communication is also autonomous in respect of actors since it is they who constitute it. However, he wrongly confines communication to text and textuality. This may work for 'auto-poiesy' but in postmodern culture communication is about, indeed primarily about, images. And because images are always linked with actions, actions share in this autonomy.

Without worrying about interpretations which may or may not be accurate, I want to use autopoietic theory to argue that the autonomy of communication (images, words, behaviour) leads to a decentring of subjectivity in the sense of a deciding or controlling institution, and to a contingency of communicative outcomes. Recognition of autonomy and its consequences (decentring and contingency) means that steering should start from the autonomy of social domains which are themselves fragmented. In that sense, the recognition of autonomy is necessary because that is how it is. This severely limits any pretensions to steering, with the result that in the actual practice of public administration, all kinds of new variants are becoming permissible. (See *e.g.* Bekkers, 1993; Godfroij & Nelissen, 1993; Kooiman, 1993.)

But in these new variants public administration is still always assumed to enjoy a hierarchically superior position, justified, whether explicitly or not, by the primacy of politics. I want to argue that public administration should take the recognition of autonomy a step further. Autonomy should no longer be regarded as an annoying and serious constraint on central steering, but should rather be postulated as a positive principle. In its steering processes, public administration should positively adopt the view:

> (...) daß die zentrifugale Dynamik funktionaler Differenzierung eine Metamorphose des Ordnungsprinzips der Gesellschaft, eine durchdringende Umstellung auf heterarchische, polyzentrische und dezentrale Arrangements autonomer Teilsysteme von Gesellschaft vorantreibt. (Willke, 1992: 7.)

And to go a step further: it is no longer a matter of functional differentiation and the existence of relatively stable sub-systems. After all, functionality still assumes some set of criteria by which differentiation can be a meaningful and, in a sense, linear process. Likewise, it is questionable whether the various sub-systems together can produce communal stability in a modernist sense. In my opinion, it is much more about a process of fragmentation which is both produced by autonomy and results in autonomy. This autonomy is distinguished by ephemerality and contingency in which administrative behaviour can only be relativist.

Administration, then, should primarily rest on a trust in autonomy. Not because autonomy self-evidently produces what is good but because confidence in autonomy is both intelligent and pleasant. It is intelligent because the administration links up with processes of social fragmentation in a flexible fashion and, being aware of the possibility of new connections, can probably stimulate them. It is pleasant because it avoids the administrative perversions (see Baakman, 1990: 234) of totalising intervention and the destruction of variety. It thereby eliminates the need for fraud, deceit and calculation on the part of autonomous actors and domains towards the central planners. This trust in autonomy is similar to what Giddens argues for.

> In larger organizational contexts, active trust depends upon a more institutional 'opening out'. The autonomy involved here can be understood in terms of responsibility and bottom-up decision-making. (Giddens in Beck, Giddens & Lash, 1994: 187)

But this is not a trust in some ideal of basic democracy or in some naïve anarchism. It is a trust based on a respect for contingency, an appreciation of fragmentation and the hope for connections.

Spontaneous Connections

It is often asserted that social fragmentation is synonymous with the disappearance of ties and connections. The romanticism and naïvety of communitarism can be traced back to this. What is actually lost are traditions as prescribed and preordained codes of meaning. Giddens speaks of a post-traditional society in which design and choice have become an obligation: a lifestyle *must* be chosen (Giddens in Beck, Giddens & Lash, 1994: 75). Here the fragments of fragmentation can be seen as constantly renewed connections, kaleidoscopic and eclectic in character, sometimes consciously designed, often arising by accident. For many, MTV is a sound-image of ephemera and fragmentation, the bane of the great musical traditions, and a chain of aesthetic kitsch. But the individual clips are anthropological jewels because in a few short minutes they reveal a complete cultural reality. Furthermore, they refer to each other, they quote from an infinite archive of styles, they assert and deny; in short they are connections and they make connections.

MTV also provides a metaphor for administrative reality. Between social domains and the organisations and actors within them, a host of horizontal connections arise. These connections can have the character of images, of communal practices, of contingent interdependency. Moreover, through organisational developments like flexible specialisation, professionalisation and contractual links, the number of horizontal relationships increases. And then there are technological developments which have an exponential effect on the horizontalisation of relationships. At the same time, these connections strengthen the autonomy of separate domains, organisations and actors because the codes of meaning attached to these connections are self-referencing. This self-reference should not be seen as absolute, as in some forms of autopoiesy, but itself as an object of fragmentation and, consequently, of more new connections.

Spontaneity, in the sense of contingency, is an important characteristic of the rise of new connections. In administration this is often seen as problematic because the future thereby becomes unpredictable and uncontrollable. One often hears talk of the unintended side-effects of administrative interventions which then have to be countered by further new administrative interventions: In 't Veld's 'law of policy accumulation'. This type of policy accumulation is a vicious circle because it aggravates the problem – spontaneity of connections – which it is supposed to resolve.

It is therefore administratively more sensible not just to allow for contingencies, for spontaneity, but actually to strengthen the preconditions for them. By aiming for conditions which allow spontaneous connections, the chance of such connections occurring can only be increased; meanwhile the conversation continues, the discussion remains open and the images are varied. In short, the autonomy of communication is respected. Through contingency and spontaneity, steering can be internalised and public administration can contribute to social plurality and the conditions for plurality.

It is clear that this demands a considerable amount of restraint from public administration. Like it or not, to trust in autonomy, to trust in spontaneity and contingency, in the expectation that connections will arise, is to marginalise public administration.

Steering at the Edges

I have often referred to meta-governance or meta-steering as a new form of steering which does greater justice to the postmodern condition of social fragmentation and the autonomy of lifestyles, practices and domains (*e.g.* Frissen, 1990; 1991; 1993b). Meta-steering can be summed up as the steering of steering. Public administration is no longer concerned with the content of specific domains but limits itself to guaranteeing the conditions for the existence of a domain (by input and/or output financing) or the structures and procedures of decision-making and image-creation within a domain. One objection to the term meta-steering is its system-theory connotations which unintentionally allow all kinds of functionalist meanings to slip in as well as the implication of a hierarchical organisation of steering actors. After all, meta-steering is usually thought of as a hierarchically superior subsystem in respect of decentralised subsystems. But neither functionalism nor hierarchical arrangement fit comfortably into our narrative of postmodern administration.

Better in this respect is the terminology employed by Bekkers. He talks about steering directed at the boundaries of organisations, and steering of the mutual dependencies between organisations (Bekkers, 1993: 98–110). Steering at the boundaries involves steering on the basis of input and output parameters with the help of incentives. Steering of mutual dependencies involves communal image-creation and communication, network patterns via relationship programs and concerted action ('konzertierte Aktion'), independent administrative agencies of the participatory type, reflexive law (or in Teubner's formulation, autopoietic law), and contracts.

Steering at the edges assumes a clearly-defined substantive position for the administration. After all, decisions have to be made as to which social activities should be protected. Rigidities and reductions of variety are introduced and, in the light of social fragmentation, become increasingly problematic and undesirable. Self-financing via insurance arrangements or direct payments seems at first the obvious solution. The next obvious step is a significant shift from the public to the private sector (which, incidentally, is much more of a hybrid than the simplistic classical free market dichotomy might suggest) and this is in progress.

Steering by interdependencies, or certainly specific variants of this, is more attractive because no substantive preferences are required of the administration. In this form of steering, the autonomy and fragmentation of domains and actors are taken as the starting point, while the conditions necessary for the emergence of connections between social domains are the point of application. In its best forms, this kind of steering is not goal- but process-oriented. It is concerned with facilitating and establishing patterns of communication, image-creation, negotiation, discussion, consultation and decision-making. The interest which becomes defined is that of independencies as such, or in other words, the interest of connections.

In my opinion, public administration should steer mainly by the processes and structures which make such interdependencies possible. Fragmentation remains a basic principle because there is no question of imposing coherence through *a priori* definition of problems and situations. Processes and structures must be designed to respect and encourage contingency and spontaneity. Creating the conditions for plurality is crucial, while the autonomy of the fragments and the domains which constitute this plurality is left untouched. The connections which can be expected are horizontal and not directed to a government perceived as holding a central position. In this image of steering, public administration focuses on the edges of social domains where both fragmentation and connections are visible and are produced. It does not intervene in those fragments and connections, but attempts to create, stimulate and follow the conditions necessary for them. Administrative steering is then no longer hierarchical in the sense of intervening in society either directly from a centre or indirectly via meta-steering. Steering moves as it were to the bottom of society, to the conditions needed for autonomy, fragmentation and connection, taking up a relativist position and honouring contingency. Steering is once again directed at infrastructures and at the social dimensions along which communication, image-creation and transactions take place. Such a conclusion is almost pre-modern.

9.3 ONCE AGAIN: INFRASTRUCTURE

The new administrative patterns discussed in this book and which in this chapter have been considered as initiating a form of postmodern administration, and the technological developments which strengthen postmodernising tendencies in these new patterns, put a classical, pre-modern, theme back on the administrative agenda: infrastructures. We can see that next to the classical infrastructure, a new infrastructure of electronic networks for information and communication is emerging, and that the modernist pattern of government intervention is being reduced to its infrastructural dimensions: steering at the edges. It should be obvious that strong interdependencies will develop between these two infrastructures.

The Archipelago

Public administration is reacting to processes of social fragmentation and is itself becoming fragmented through the internal dynamic of 'bureau-politics'. We can see it in the relationships between layers of the administration where localisation occurs as a consequence of decentralisation, of the struggle for autonomy by lower units, or of independent regional links which can also develop an international dimension. We see it in processes of autonomisation where, through internal independence, functionalisation, farming out and privatisation, monolithic government organisations start to break up. We see it in the processes of horizontalisation in the relationships between government organisations, independent units and social actors where temporary arrangements are constantly being made within networks and via contractual relationships.

The fragmentation of the administration is further promoted by new forms of steering and policy-making. A common characteristic is greater interactivity between administrative and social actors. This weakens the linear process of policy-making whereby the administration ordains, makes plans, consults, invites participation and gives orders. Interactive steering in policy networks means that there are no dominant actors and the initiative does not necessarily lie with the administration. Different parts of the administration operate in different policy networks, each of which will have its own particular culture. That too leads to fragmentation.

Fragmentation of organisation and fragmentation in the policy process together lead to a wide proliferation of images and image-creation processes. This proliferation incidentally was already a reason for fragmentation in the administration. Establishing clear-cut codes of meaning and significance is becoming increasingly problematic.

There are of course counter-movements. Impassioned pleas are made for restoring the primacy of politics. The spread of independent agencies must be moderated and controlled (Commissie-Sint, 1994; Algemene Rekenkamer, 1995). A General Administrative Service is being set up (which seems to be primarily concerned with protecting its own interests). The steering powers of core departments ought to be strengthened (Van Twist & In 't Veld, 1995).

But these counter-movements are in vain. They are all pervaded with the modernist obsession with control, they define a solution for a problem which is itself a set of solutions for a problem – boundaries to central steering – which is being formulated as a solution. Once again the past lessons of failed attempts to combat fragmentation by central co-ordination are being ignored. Social fragmentation is being defined as a problem instead of being adopted as a basic principle.

Society has acquired the character of an archipelago, and public administration is following suit. Public administration can, it is true, be regarded as self-referential, but the self-reference exists mainly in patterns of communication with social domains, through images, text, and interventions. And because of the relative autonomy of this communication, patterns of fragmentation spread through the administration. Connected with this is the deliberate re-design of public administration. Public administration too is becoming an archipelago.

Of course, the archipelago is a metaphor. It symbolises the following aspects:

- Public administration is fragmenting organisationally through decentralisation, independent agencies, functionalism. The (relative) number of independent agencies with public responsibilities is increasing.
- The autonomy of separate organisational parts in the administration is growing, through professionalisation on the one hand, and horizontal network connections on the other.
- The structural and cultural differentiation of policy networks increases the fragmentation in operational and image-making practices in the administration.
- Connections between parts of the administration and social domains are becoming more contingent; institutional permanence can thus rarely be achieved.
- The fragmentation has far less to do with the differentiation of subsystems than with the fragmentation of 'aspect systems'. Connections should not be seen primarily as a relationship pattern between subsystems, but more as a pattern of temporary aspectual inclusion in relation to operation and image-creation.
- The archipelago is heavily layered and aspectually dimensioned. The relationship between the layers and the dimensions is neither hierarchical nor an integrated whole which gives significance to its parts. Rather, the whole is contingent and always temporary. It is an aspect. In fact, the whole is rather less than more than the sum of its parts.
- The connections between the parts of the archipelago are primarily forms of information and communication, both at the level of image-creation and operation and at the level of infrastructural provision.

It is precisely in respect of information and communication that a technological infrastructure comes about. This infrastructure has particular politico-administrative significance because, by promoting fragmentation, it strengthens the connections in the archipelago. Postmodern administrative theory must face up to this.

Information and Communication: Cyberspace

The connections in the archipelago are concerned with information and communication. In the first place, this involves creating images and developing communal identities for and by actors and organisations. In the second place, it involves horizontal organisational networks. Thirdly, it involves processes of policy-making and decision-making. As has been repeatedly pointed out, in the actuality of postmodern administration, fragmentation and connections can not be separated, and contingency must be accepted and pursued.

Technological developments in the communication and information infrastructure are widely known as digital highways and much has been written about it (*e.g.* Bekkers, 1994; Berenschot, 1995). The digital highways are made up of infrastructures of cable and satellite links, networks and networked connections, communication standards and an explosion of applications, providers and services. The Internet is a part of this. And it is important to realise that it is not a single physical infrastructure, but an expanding complex of connections, possibilities and agreements which is not only physical but also consists of processes and procedures. Furthermore, it is not just a physical tool and a high-capacity external instrument, but a technological realisation of theoretical and intellectual capacities of which humans are a part and which increasingly determines human behaviour and substitutes for it.

It would be naïve to assume that this infrastructure will produce 'the good society'. Bill Gates might hire the Rolling Stones merely to promote a new product, but these icons of Rock 'n Roll revolt were also bought to strengthen a monopolistic conglomerate. Violence and oppression, too, can find untraceable routes around the digital highways.

The administrative importance of the digital highways is unmistakable. In the first place, they intensify fragmentation and contingency. As a result, public administration itself becomes more fragmented and contingent. Secondly, information-provision and communication between the administration and social domains will make increasing use of this infrastructure, quite spontaneously but also as a result of deliberate redesigning within organisations and the administration. Future generations, moreover, will take it entirely for granted. Thirdly, this infrastructure reinforces the process of virtualisation which is altering organisation patterns and steering processes and which is also giving an autonomous technological impulse to the proliferation of images and image-creation. In the fourth place, the notion of territory is being affected by the dual effect of glocalisation. The territorial basis of public administration is losing importance and this is tied up with such developments as functionalisation and independent agencies. The infrastructure of the digital highways is a cyberspace which is almost unmanageable and whose development is highly contingent. The connections making up this cyberspace have a strongly virtual character. Rheingold (1994) has discussed the virtual communities of the Internet; Mowshowitz (1992) points out the virtualising of economic production processes and relationships, and the need for politics and administration to adjust to it. Bellamy and Taylor (1994) describe this adjustment which, though partly spontaneous, will also need a deliberate redesigning as the 'Information Polity': politics and public administration are acquiring the characteristics of an information and communications infrastructure.

The infrastructural change of direction which postmodern public administration will have to make in response to social fragmentation and contingent connections, is not only made unavoidable by the development towards cyberspace, but can also take form through that cyberspace. Seen from cyberspace, the proposed postmodern reorientation of public administration is not just inevitable, it is also the only way to avoid obsolescence. The classical tendency to increase government intervention in the digital highways – a combination of regulation, stimulation and instrumentalising in the interest of established goals – will be ineffective because it is diametrically opposed to the central characteristics of the technology. Digital highways will undermine that classical orientation by means of the virtualisation, fragmentation and contingency which they produce and on which they are based. Administrative effectiveness can only be achieved by acceptance and adapting to reality.

At the same time, it is precisely in this cyberspace that the new postmodern orientation of public administration can take shape. I am not referring so much to the various proposals to use ICT and the digital highways in the interests of the administration by improving its communication with the public (Depla, 1995; Ministerie van Binnenlandse Zaken, 1995; Schalken & Flint, 1995). Of course, such initiatives are welcome and will stimulate much-needed infrastructural reform. More important is that in a number of respects cyberspace should be seen as an administrative infrastructure in its own right. The existence of that infrastructure is a crucial condition for patterns of communication and information, and those patterns of communication and information are not the outcome of an infrastructure but in an important way coincide with it. Cyberspace *is* the postmodern administration which is both fragmenting and connecting and which follows and emulates the fragmentation and connections of social domains.

From Scale to Scope and Meaning

A fundamental consequence of the digital highways for administration is the reduction in scale. In Lash's words:

> We extend such an idea of 'information and communication structures' from the level of the firm to entire production systems and then to the flows of information and communications (and immigration and tourism) that are taking place on an increasingly global level. Thus we are also talking about economies of space, which at the same time are extensively globalized and intensively localized. These information and communication structures, which stretch over broad areas of space and compress time, contain not just informational signs but also images, narratives and sounds – that is, aesthetic or hermeneutic signs. (Lash in Beck, Giddens & Lash, 1994: 213)

Reduction of scale is a technological option. Scaling up without becoming large-scale is operationally also possible. That means that scale is losing its importance in the design of organisations and processes. Furthermore, communication and information can pass via the digital highways with virtually no space–time limitation. Not only does this have economic and social consequences, but also and in particular it is culturally important because identities and lifestyles cease to be tied to territory and codes of meaning become fragmented. The consequences for the administration are fundamental. For a start, the importance of the national state is clearly being eroded by economic, social and cultural globalisation. (See also Crook *et al.*, 1992: 101 ff.) The integrity of the national state in generating economic policy, in providing social security and as the source of a national cultural identity is disintegrating. ICT and digital highways are important causes and forms of this.

But of even greater importance is the disappearance of scale as the rationale and legitimisation for the design and functioning of the administration. Its importance lies in the decentring of organisation patterns through virtualisation and horizontalisation, in the decentring of decision-making processes through the proliferation of images and image-creation, and through the contingency of social and technological developments as well as in the multi-layered archipelago structure with its aspectual connections.

So administration can now be designed on the basis of scope and meaning. That is to say, it is precisely the infrastructure of cyberspace which has made it possible to organise the forms of information and communication to include as many actors, organisations, image-creation processes and connections as possible. In principle there are practically no limits to scope, while at the same time it is possible to strive for maximum fragmentation and contingency of images. Cyberspace does not need such patterns to be fixed or congealed to make them work; rather the reverse. Ephemerality can also be realised if so desired. Yet scope and meaning do not have to be defined *a priori* because they already exist as realisable possibilities within the infrastructure.

The multi-layering of administration – sufficient for fragmentation and connections, as for their aspectual character – is unlimited and in postmodern perspective must be so. The administration then acquires the characteristics of a rhizome as well as considerable fluidity. Its coherence is changing and contingent and must be kept that way so as to avoid destroying fragmentation and to allow connections to be pluralistic.

The different coherences which continuously and temporarily arise have no hierarchical relationship with the fragmentation which they hold together. The idea of a centre disappears, just as in cyberspace there is no centre. That is an important principle for postmodern administration. One might call it a style feature.

9.4 FORMS AND STYLES

Postmodern administration is primarily a question of form and style. Of form because steering is becoming less a question of content and more procedural and structural. Of style because technocratic intelligence and aesthetics are becoming important motives.

Postmodern administration consists on the one hand in the design of administrative structures, processes and arrangements, and on the other in an orientation towards forms of social self-steering and self-organisation. The design of administrative structures, processes and arrangements is necessary because of social fragmentation. The operation of public administration, precisely because of the self-referentiality of the communication patterns into which it is woven, is necessarily fragmented and therefore fragmenting. Issues of organisational structure continue to need attention even though, paradoxically, the importance of organisation structures is declining (see also Peters, 1992). These structures are of course relative because they are primarily focused on flexibility and adaptability. (It is analogous to flexible specialisation in industry and the resultant need for permanent re-engineering.) Herein lies the link with style because flexibility and adaptability, while obviously requiring a thorough understanding of organisational structures, are primarily a matter of attitude and cultural orientation.

In parallel to the structuring process, the procedural dimension of steering is also becoming increasingly important. Because of fragmentation and contingency, policy-making and image-creation assumes that public administration will define its role procedurally: not oriented toward outcomes but toward openness and accessibility and the plurality of actors and images. Shaping these processes, and especially protecting their fragmented and contingent quality, is an important task if plurality is as inevitable and necessary as it is desirable. With Tjeenk Willink (in Tjeenk Willink, Derksen, Oosting & Nekkers, 1991) we can describe this as the director's function of public administration so long as we do not imply a director who supervises and has a deterministic image of the outcome. Public administration will certainly seek connections, but without knowing what they will be: wanting to connect means having to be fragmented.

In structuring, policy-making and image-creation, public administration needs to be orchestrated or 'arranged'. This is far from saying that the administration is also the 'arranger'. Restraint is necessary because many 'arrangements' result from processes involving many participants, many images and contingent outcomes. However, public administration can pursue permanence in its operations by expressing confidence in the autonomy of social actors and processes. Once again we see a style feature. This seems to me to be a more convincing motive for the pursuit of permanence than Salet's use of the intrinsic and the normative. When he discusses institutionalising norms of behaviour, he has to resort to hierarchical relationships (Salet, 1994: 101, 124 ff.).

Design by the administration therefore relates to processes of information and communication. Moreover, information and communication become fragmented in their progress and outcome by the relative autonomy of their related technologies and through postmodernising processes. Contingency is the result. Fragmentation, however, is also a pattern of connections, not as a totalising final outcome or the result of an *a priori* overview, but as a form of plurality.

Design by the administration is then infrastructural in a dual sense. Firstly, the administrative and electronic infrastructures for information and communication in and between social domains are the intervention points for steering. Secondly, the infrastructures themselves, with their relatively anarchistic and contingent character, are becoming the primary structures and processes of steering. Public administration is not only a question of design. It is primarily form coming into being.

Design may be important, but so is style. One of the hallmarks of postmodernisation is its preoccupation with style and style features which are as much its cause as its effect. We discussed this earlier in our treatment of mass culture and lifestyles. Many writers have noted the postmodern attachment to eclecticism, quotation and pastiche for which history, tradition and narratives constitute a freely accessible archive (Crook, *et al.*, 1992; Harvey, 1989: 44).

Because there are no more integrating criteria by which behaviour and practices can acquire meta-narrative legitimacy, there has been an 'escape' into externals and design. That is also true of public administration, which faced by postmodernisation, opts for a postmodern position. Since public administration can no longer demand a preferred outcome for structuring, policy-making and image-creation without resorting to a totalising grand narrative and destroying social fragmentation, there remains nothing else than aesthetics; an aesthetics of intelligence and beauty. Public administration will have to be intelligent. It must accept the principles – if we may use that term at all – of fragmentation and contingency. And by that I mean active acceptance in the sense of a constant readiness to acknowledge fragmentation and contingency, and the freeing of creativity and talent in order to achieve it. That applies even more where technological developments focus increasingly on knowing and therefore – paradoxically – increase ignorance and create a permanent state of 'don't yet know' (*e.g.* Giddens, 1990: 176.). Willke draws the conclusion that the most pressing task of the state is 'the provision of a knowledge-based infrastructure', not in order to monopolise knowing but, constantly to produce 'anti-knowing' in the public interest (Willke, 1992: 267). He seems here to imply the possibility of 'better-knowing'; a position which I can not share. I would much rather take the principle of not-knowing as the starting point for postmodern administration. Only in that way can fragmentation and contingency, and especially their connecting qualities, be respected.

Intelligence as a style feature is in some sense a technocratic choice. Not in the classical meaning of the word which Fischer (1990) criticises for its totalitarian claim to truth, but rather in a postmodern interpretation which is cautious and modest just because it recognises the fragmentation and contingency of knowledge and thus prizes it the more.

> Gegenüber den Zumutungen und Anmaßungen einer umfassenden Steuerungsfunktion der Politik (...) kommt es für die Staatstheorie darauf an, die Politik mit einem geeigneten Modell ihrer selbst vor einer Überforderung durch Hyperaktivität und Allzuständigkeit zu schützen. Dies verlangt die thematische Verortung von *Selbstbindung* an Standards der Respektierung der Autonomie gesellschaftlicher Teilsysteme, *Selbstbegrenzung* im Sinne einer Fähigkeit zur Reflexion der Intergrationbedingungen einer komplexen defferenzierten Gesellschaft; und es erfordert eine neue *Bescheidenheit* der Politik aus Einsicht in die Notwendigkeit dezentraler Entscheidungsfindung in undurchsichtigen und risikoreichen Problemfeldern. (Willke, 1992: 316)

Willke describes such a position for politics and public administration as 'ironic', following Rorty, who incidentally reserves the term for the private domain. Irony then is a quality of public administration because it does not imply the complete withdrawal which would follow from scepticism, but on the contrary expects activity from the administration.

An ironic position is also necessary for reasons of 'beauty'. Beauty is the second dimension of the administrative aesthetic which I am describing and consists primarily in a kind of hedonism: administrative infrastructures must aim not only to produce intelligence but also to avoid cruelty. That means in the first place that they aim to prevent domination and oppression, violence and aggression. That is, of course, a classical liberal creed because it puts the freedom of actors and processes first, and judges interventions, either by actors or resulting from processes, by the degree to which they destroy freedom. But in the second place, the avoidance of cruelty means that administrative infrastructures have to be open; they must do justice to fragmentation and contingency and they must defend and stimulate conditions for fragmentation and contingency. If there is a highest goal, it is the agreement that plurality is always better than uniformity. The possibility that a grand narrative might fill the empty place of power must be resisted passionately. And for that too an ironic position is necessary. After all, such an orientation requires the intelligence of not-knowing as well as a respect for and confidence in knowing-otherwise. Postmodern administration must fully adopt such an ironic position just because its social position is still potentially dangerous.

Administrative theory must tell narratives about this ironic position.

9.5 POSTMODERN ADMINISTRATIVE THEORY

In many respects, the narrative which I am now telling conflicts with narratives I have told in the past. From a postmodernist perspective, there is nothing wrong with that. But it is striking that the word 'must' has been used so much in my description of public administration and its consequences for administrative theory. However, that is merely a rhetorical strategy since, after all, nothing *has* to happen. What I am trying to indicate are the consequences which I see when my story about social postmodernisation is combined with the structuring and functioning of public administration. Underlying it is a possibly modernist conviction that public administration must adapt intelligently to social developments and make connections. It is partly a question of survival. Like Ringeling (1993), though for quite different reasons, I am greatly attached to public administration, certainly the Dutch, because it has a long tradition of postmodern behaviour *avant la lettre*. It possesses a high level of creativity and large amount of professional talent.

My narrative, then, should be read as a series of proposals (made more urgent by the use of 'must') directed at the administration. Or preferably, an invitation. It is an invitation which does not have to be accepted, and certainly not if the narrative is rejected on aesthetic grounds.

Cool Science

I have argued previously that administrative theory ought to be a 'cool' science (Frissen, 1989; Frissen, 1991). Description and analysis should be placed first, since prescription is not a scientific but a strategic activity. But it might still be insisted that a strong theoretical orientation towards problem solving, improvement and change is problematic: methodologically since social sciences can not be predictive because of their reflexivity and contingency; normatively because the good and the better can not be defined scientifically. If fragmentation and contingency are important characteristics of postmodernisation, a science such as public administration can only bear witness to it and adopt a very modest position when giving advice or making recommendations. 'Wisdom is perhaps also ephemeral' (In 't Veld, 1984: 92).

A postmodern discipline of public administration remains and must remain a cool science. There are various arguments which can be put forward for this. In the first place, there is the condition of not-knowing which is intensifying, particularly now that the infrastructure is becoming increasingly knowledge-based. Administrative theory produces knowledge and insight which show that knowledge and insight are relative – thus one might sketch the spiral of reflexivity. The relative nature of every prescription is thereby heightened.

In the second place, fragmentation and contingency undermine the possibility of Wildavsky's classic adage that 'speaking truth to power' is public administration's most important task. Ringeling's argument that academics and theorists should discuss the question of what constitutes 'good policy' (Ringeling, 1993: 303) is, in my opinion, problematic. It is obvious that a 'cool' science of public administration can not be regarded primarily as a critical discipline which judges policies and unmasks intentions.

Thirdly, the coolness of science is preferable to the warmth of passion if the latter is concerned with the grand narratives of the good state and scientific truth. Public administration can produce no convincing argument as a normative grounding for steering and policy-making. It is enough to rely on the fact, which Rorty insists repeatedly is ethnocentric, that a liberal state is the most agreeable one. Nor can public administration claim that it wants to speak scientific truth. Every utterance is, after all, contextual (Van Twist, 1995). Truth can not be found outside its representation – any science of public administration is therefore metaphorical. Wanting to know better is a dangerous pursuit.

Fourthly, the postmodern discipline of public administration is cool because it is disinterested. That is to say, the interest which it serves can only be to articulate interests which are as highly fragmented as possible. Steering and policy-making must be as open as possible in order to avoid any *a priori* definition of exclusion. That is why metaphors are important because their use is evidence of a fundamentally relativist position. Only thus can plurality be respected without public administration claiming the ability to conceive of the substance of this plurality.

> It is questionable whether the ideal of a literal, non-figurative, scientific language which describes things 'as they are' is at all feasible. Certainly no such language yet exists and attempts to develop such a language have all failed miserably. The use of metaphors in science seems unavoidable. In my view, the question is not so

much how to avoid metaphors, but rather how to make their use, which is as risky as it is unavoidable, more manageable. (Van Twist, 1995: 47)

The answer to that question must be sought for in irony.

Ironic Reflection

A cool science is convinced of the relativity of all knowledge, it no longer pursues *the* truth, it denies the possibility of a 'final' grounding, and it is disinterested. It likes ambiguities, dilemmas and paradoxes. There is nothing shocking about that. Even a writer like Hoogerwerf (1995) testifies to the need to think in dilemmas and might therefore be described as postmodern were it not that he postulates a number of deeper-lying values to counterbalance the dilemmas.

Cool science, however, need not be detached or distant. In that sense, I should like to modify what I have repeatedly claimed about the position of theory in respect of its object, the administration. Van Twist's conclusion that by so doing I end up agreeing with Hoogerwerf (Van Twist, 1995: 32, note 8) is surprising but true. It is by remaining cool that administrative theory is able to plunge into the arguments and practices of public administration, both through research and public debate. I would describe the orientation which it must adopt for this as 'ironic reflection'.

I have borrowed the concept of reflection from what Beck, Giddens and Lash (1994) call 'reflexivity' (see also Giddens, 1990). 'Reflexivity' expresses a number of things:

- Society's confrontation with unintended effects which can not be processed by the institutional standards of that society;
- the condition of non-knowing, of unpredictability, of contingency which is the consequence of the increasingly prominent place accorded to knowing within society;
- the growing proliferation of images and meanings resulting as much from interferences between them as from the image culture.

The discipline of public administration has the important task of making this reflexivity visible. That involves attempting to analyse social fragmentation, to trace the conditions under which it exists and then trying to reflect this fragmentation - naturally in fragments and metaphor. In advising the administration and participating in public debate, postmodern public administration has to clarify the reflexivity of the different social domains as well as the administration's contribution towards that reflexivity. It is impossible to do that at the level of content. At the very most, the science of public administration can make available to administrators some relevant insights into reflexivity which other disciplines, or more often journalism, literature and the arts, have obtained, and draw lessons for the design of structures and steering processes.

Public administration then must translate the conditions for fragmentation and contingency into proposals for setting up structures and steering processes which do justice to these conditions in creating an administrative infrastructure which follows them. This in turn can be seen as another condition for fragmentation and contingency.

At the same time, this reflective task must be ironic. With Willke I disagree with Rorty's assertion that irony can only be a private virtue. The fact that vocabularies are idiosyncratic, highly contingent and changeable is, in my opinion, equally true *a fortiori* of social vocabularies. Fragmentation (in Willke's terminology, 'Differenz') is an important characteristic of social domains. Connections exist between them; to put it even more strongly, fragmentation can be seen as a form of contingent connections. It is precisely the ability to see these connections as well as their relativity which distinguishes irony from scepticism and tragedy (Willke, 1992: 320–321).

What Rorty regards as the only desirable goal of liberal society – the avoidance of cruelty – demands an ironic standpoint in both the practice and theory of public administration.

> Insofern hat Liberalität es genau mit einer bestimmten Qualität der Kontrolle systemischer Kontingenzen zu tun: einer Rationalität der wechselseitigen Akkordierung von Kontingenzen, die dem Ziel folgt, die Autonomie und prinzipielle Kontingenz der Akteure zu erhalten und zugleich in die Aktualisierung der Kontingenzen das gemeinsame Kriterium de *Schadensminimierung* einzubauen. (Willke, 1992: 323)

The problem with Willke's conclusions from his argument, is that he wants to translate this ironic position back into functionalist terms of problem-solving and 'contingency control' within and between highly complex social subsystems. He then attaches the normative conclusion that the administration can be seen:

> (...) als 'lokaler Held' in seiner Funktion der Bindung territorialer Solidaritäten und als Anknüpfungspunkt für die Produktion territorial eingegrenzter kollektiven Güter. (Willke, 1992: 371)

In my opinion, the ironic position of the theorist who wants to reflect the conditions for fragmentation and contingency in the interests of the administration, means that functionalist and normative reasoning must be abandoned, particularly as a justification. Rorty's notion of irony has practical and theoretical significance for administration in so far as it leads to theories and practices which are more intelligent and attractive. More intelligent because they express in words and deeds that nothing more than not-knowing is possible; more attractive because they seek a libertarian plurality which is self-sufficient and requires no external justification.

Technocracy as Art

Narratives such as this are frequently criticised for being technocratic (*e.g.* In 't Veld in Tjeenk Willink *et al.*, 1994: 104). It is a criticism because in virtually all debate, 'technocracy' is a pejorative term. It refers to the power of experts who are not subject to any controls, either political or social. Technocrats are those who want to replace the irrational, especially in politics, by rational expertise, as functionally more effective and therefore in a normative sense as bringing us closer to the 'good society' (Fischer, 1990: 22).

However, I want to argue in favour of technocracy. For me the term has a different meaning from that of most critics. It still involves expertise, knowledge and insight, understanding but in an ironic variant. It is, moreover, an expertise which is not one-sidedly scientific and positivist and which does justice to the richly variegated technocracy which constitutes administration. It includes a much clearer recognition and acknowledgement of fragmentation and contingency than that implied by the prevailing image of engineering rationality and drawing-board policy making. In my experience, scientific expertise is being called on more and more frequently to provide clearer insights into fragmentation and contingency in order to design structures, processes and arrangements which will reflect such fragmentation and contingency.

The technocracy which I want to defend is convinced about not being able to know precisely because in a postmodern society, communication- and information-infrastructures heighten uncertainty and complexity. That technocracy attempts to trace the conditions for uncertainty and complexity, not for better control over the contingencies to which they give rise but to understand them as preconditions for steering and to include them as constituent conditions in information and communication infrastructures.

It is a technocracy which sees the aesthetics of intelligence and beauty as a more convincing guideline than the rationalistic implementation of the political will on the basis of a grand narrative. In its irony, this technocracy is anti-totalising and knows that it only deals in metaphors and that its narratives consist of proposals and invitations. Within this aesthetic orientation, technocracy – and postmodern administrative theory – is a form of art. It does not strive for truth but for pleasure.

> We love doubts, questions and above all we love pleasure. Your body is what you are, and all knowledge begins with desire. And for what else than pleasure? (De Moor, 1993: 29)

The postmodern theorist of public administration attempts to support technocracy with the help of ironic reflection. That is the 'redescription' of which Rorty speaks: 'The liberal ironist just wants our *chances of being kind, of avoiding the humiliation of others*, to be expanded by redescription.' (Rorty, 1989: 91)

That is simultaneously modest and ambitious. Its modesty requires a systematic refusal to pursue the just, the good or the true or to make them the basic principles for the theory and practice of public administration. What its ambition means for the theory and practice of public administration has been the subject of this chapter. There now only remain politics and political theory. It will become apparent that they too are affected by postmodernism's ambitious modesty.

10 Politics without Properties

In the previous chapter about fragmenting and connecting governance, politics was not explicitly discussed even though it was always implicitly present. After all, a diagnosis of fragmentation and contingency leading to horizontalising and virtualisation in the administration is bound to be problematic for politics. When, for reasons of intelligence and aesthetics, a plea is made for a fragmenting and connecting governance, implying the necessity and desirability of a relativist position, the problems for politics become even more acute.

The decentring of subjectivity also affects the position of politics, both empirically and normatively. Empirically it undermines the hierarchical pyramid of organisation; normatively it weakens the grounding and unifying role of meta-narrative. Postmodernisation thus affects politics in both substance and form. At the theoretical level it brings in normative political philosophy as well as empirical political science. In practice it is about the legitimacy and design of politics.

This chapter brings us to the end of my narrative. This suggests that we have reached the heart, the moral, of the story. In a way that is true since modernist politics sees itself as the decision-making centre of society (in the sense of Easton's 'authoritative distribution of values'), as representing the subjectivity of the popular will, and as defining and grounding morality. It is that self-perception that postmodernisation has eroded and which postmodernism regards as unfeasible and dangerous.

At the same time, the place of this chapter might symbolise the fact that politics too has reached an end. That also is true to the degree that it will be argued that the traditional, modernist theory of politics has become obsolete. But it will not lead to the conclusion that politics is disappearing from postmodern society. So there is some hope for those who might regret the passing away of politics, even though that will have to be understood in ironic terms. I shall attempt here to indicate what postmodern politics means and how political postmodernism might be formulated. The following quotation provides a foretaste:

A realistic conception of democracy should in fact abjure any definition of 'public ethics' used, as in Rawls, to design a 'public' anthropological model, to select needs, establish common values and found universal rights. It should, on the contrary, recognize its own limitations and its own radical contingency and leave to other social spheres – culture, art, music, friendship, love, scientific study, and even religious belief – the enquiry into ultimate ends and the promotion of values. (Zolo, 1992: 180.)

10.1   POLITICAL THEORY: A MODERN PROJECT

In recent decades, after a long period of positivist domination, political theory has enjoyed an unprecedented revival. This is reflected in the work of Nozick, Rawls, MacIntyre which has led to wide debate, in the never-ending production of Habermas, the 'last of the Frankfurters', and in the recent resurgence of communitarism and republicanism. This revival can be explained as a reaction against narrow-minded positivist political analysis and interpretation. On the other hand, it is just as much a reaction to the process of social and political postmodernisation. In this connection it is striking that Hoogwerf's Dutch variant of 'In Defense of Politics' distances himself quite explicitly as much from the 'public choice' approach to politics as from postmodernism (Hoogerwerf, 1995). It is not for nothing that Habermas writes about the uncompleted project of modernity. It is in this incompleteness that the need for normative politics still lies.

Futile Romanticism 2

It is the dream of normative political theory to formulate a universal theory of the good society. Justice and rationality constitute important concepts in the theory, while the republic of citizens and the legitimate community are the political forms in which justice and rationality can be effectuated.

That dream is more particularised in modernist political theory because it is based on a conviction that modernity is capable of realising all the ideals of the Enlightenment. Those ideals imply that humans are potentially capable of fulfilling their subjective natures and exercising mastery over contingency. Democratic politics are an expression of this belief. The fulfilment of human subjectivity is taken to mean emancipation because the realising of freedom is inherent in the fulfilment of subjectivity. This freedom presupposes and makes possible the fundamental equality of mankind which is rooted in the fact that every human is endowed with reason. No-one may be excluded and any attempt at exclusion is a betrayal of the promise of emancipation and justice.

The Enlightenment is therefore both the driving-force and the touchstone of progress. The continuous development of human reason will inevitably increase human freedom. The development of reason and the growth of freedom will take us closer to the 'good society'. And the good society is defined in terms of fulfilled subjectivity and maximum rationality.

But history does not abide by this teleological scheme. The development of reason has resulted in the domination of functional rationality, which has either reduced any substantive-rational foundation to trivial abstraction or has broken it up into localised or fragmented codes of meaning.

Even more of a problem is that progress has not in any way led to the 'good'. The drive to achieve the dominance of reason has become as discredited as it has been successful. It is manifest in the bureaucratic rationality of this century's genocides; it is insidious in what Beck, Giddens and Lash (1994) call the contingencies of reflexivity: the unmanageability of unintended and contradictory consequences.

And if that were not bad enough, it is precisely this progress, with the knowledge and technologies which it has produced, which have undermined the notion of subjectivity. The subject is being deconstructed and is vanishing from the centre of the historical process. Neither can the externalising of subjectivity in politics any longer be grounded in universal notions of justice and freedom.

So the historical contingencies of progress should be defined in political theory as a 'betrayal of reason' (Stuurman, 1986: 31). And there is a frantic search for a constituting narrative that can reformulate 'pure' and just reason. But as Stuurman rightly points out, every attempt to tell such a narrative represents a political and moral choice. And the history of modernity has made every political and moral choice an arbitrary one.

The fact that choices are arbitrary is only a problem for those who seek a foundation for morality and politics and believe in its possibility and necessity. Their desire and their belief, however, can not escape from the fragmentation and contingency produced by the modernising process which have led to the qualitative break of postmodernisation.

Clinging to the conviction that a normative political theory can provide a universal foundation for justice and the good society, is a form of romanticism. That kind of political theory dreams of harmony and unity and in its modernist guise believes that reason can produce this harmonious order. Such romanticism is futile because reason and its realisation in knowledge and technology are themselves a cause of fragmentation and contingency. So harmony and unity have necessarily become local fragments which might be able to make connections but which have no coherence or foundation.

Dangerous Idealism

But surely there can be no objection to romanticism? Not at all, so long as it is futile. We can then smile indulgently, and regard it as an interesting phenomenon and a specific lifestyle. We can begrudge no-one their idealism. Romanticism can be a form of hope, the expression of utopian yearning, a narrative of desires. Everyone is free to hope, and one can take aesthetic pleasure from the narratives.

But the romantics themselves will not let us get away with it so easily. They are persistent and pretentious. They will object firmly to the epithet 'romantic'. They do not see their narrative as *a* narrative, but as *the* narrative. At least, their contribution to any debate is to ensure that the outcome is their narrative and no-one else's. Or, like Habermas, they try to formulate conditions under which the debate can be conducted rationally: communicative rationality. But even Habermas's apparently modest position is pretentious because for him the conditions for a power-free dialogue imply the universal norm of an ideal power-free dialogue (Stuurman, 1986: 31; Habermas, 1987). Reasonableness is not only a convenient instrument, but also the inescapable ultimate goal of modernity. And wherever historical contingencies seem likely to produce the opposite of reasonableness, through the reflexivity of reason, the subject must introduce the norm of emancipating reason in order to realise his subjectivity.

But even the persistence and the pretensions of the romantics could be shrugged off if it were not for political romanticism. Pretensions and persistence can become dangerous when the political domain gets involved.

Politics is concerned with the collective, it is equipped with monopolies and it is the historical embodiment of the pursuit of subjectivity. Politics sanctions the foundations of law and is thereby both immanent and transcendent in society. (See Foqué, 1992: 6) It is that particular combination of immanence and transcendence which makes up the institutional character of politics and which makes the pretensions of modernist political theory so dangerous.

Modernist political theory, however friendly, noble, modest and idealistic it tries to be, ultimately wants to recount a grand narrative and because it believes that the narrative will be conclusive it wants politics to be founded upon it. Whoever aims to define the good society philosophically and believes that the political system can create the practical conditions for it, may in specific historical circumstances be subversive because he will criticise the existing hegemony; but ultimately he is dangerous. The danger lies in the fact that any pursuit of a grounding which does not regard itself as 'contingent' but as the expression of universal validity is fundamentalist. If it did see itself as contingent, it would automatically be particularistic and conceive of itself as but one of a number of possible narratives. Bovens suggests that a way out of this contradiction between particularism and universalism is to accept it as part of the *condition humaine* and recognise both as valid markers in the debate about political principles (Bovens, 1988: 352). I can not go along with this. For if we reduce the contradiction to a condition of existence, we inevitably recognise universalism as a pursuit, as a sentiment. And as such it becomes particularistic.

The fundamentalism of the grand narrative is dangerous if it tries to legitimise the struggle for political power. It then becomes inextricably tied up with strategies of inclusion and exclusion. The denial of the universality of the principles formulated by normative political theory can never itself become one of those principles. It would only be possible for a political theory which accepted contingency and characterised each validity as a temporary construction or arrangement. Normative political theory aims specifically at 'the triumph over history and contingency' (Ankersmit, 1994b: 134). The danger is that it inevitably leads to:

> (…) a political attitude which will lead you to think that there is some social goal more important than avoiding cruelty. (Rorty, 1989: 65)

And because politics carries with it compulsion and obligation, romanticism and idealism are far from innocent if they persist in claiming that they can give politics a normative foundation.

Totalitarian Normativity

I still have not sufficiently blackened the vocabulary of normative politics. Because it tries to ground the normativity of politics, it is also totalitarian. For if it seeks to ground politics in the universal validity of a specific normativity, and sees politics as the realisation of that normativity or of the conditions required to validate that normativity, it is totalitarian. It is claiming not only to know which norms are universally valid, but considers it self-evident that it is legitimate to pursue empirically the concretisation of that universal validity through politics. That is why, for instance, Christian-Democratic communitarianism, notwithstanding the very modest role which it attributes to politics, is nevertheless totalitarian because of the normative connection between politics and morality.

A normative connection between politics and morality is totalitarian because it may lead to, indeed sometimes desire, the domination of a single political morality – however tolerant of pluriformity it tries to be (see *e.g.* Hirsch Ballin, 1994b) – and because it legitimises the ever-present totalising potential of political morality. The empirical contingency of each concrete political situation is neutralised by the unshakeable knowledge of a political morality. In modernity there always arises a connection between the functional rationality of the pursuit of domination and the belief in universally valid normativity. The totalitarian danger of this scarcely needs pointing out in the light of the historical evidence.

Even if claims to universal normative foundations are wrapped in the friendly terminology of justice, liberty and emancipation, and even if its reasonableness is made central, the concrete political outcome remains totalitarian. For the inevitable result is a state which moralises and which legitimises its policies on the grounds that they enhance the moral content of society:

> Restore responsibility for the transfer of norms and values to families, schools and churches. The government must tune its policies to this and underpin its trust in family relationships by, for instance, adjusting the fiscal and social system to communal responsibility and capacity. It must support private initiative in, among other things, education, health and social services (…) (Hirsch Ballin, 1994b: 85)

That is totalising politics because it tries to combat fragmentation and contingency with a narrative of necessary moral foundations. There may be political arguments about its universal validity but even in disagreement it can not be defined relativistically. Right and might automatically become absolute variables which are morally legitimised. Fragmentation and contingency can then only be regarded as phenomena to be resisted. It is these dangers of totalitarian normativity, associated with the romantic idealism of (modern) normative political theory with which postmodernism takes issue. In order to articulate the postmodernist attitude to politics I need first to define, mainly theoretically, the political significance of postmodernising processes, what the content of politics might be and what tasks confront a postmodern political system.

10.2 FRAGMENTATION, CONTINGENCY, INDIFFERENCE

The pretensions and ambitions of normative political theory have their counterparts in actual politics. Its pretensions and ambitions have a normative component: the principled position of politics in modern societies, and an empirical component: the structural position and the functions of politics in modern societies. Postmodernising processes have been problematic for politics in both respects.

Before exploring this further, I should say a few words about concepts. When I speak of 'politics', I am using an extremely vague term. I do that deliberately because politics is a vague phenomenon which is at the same time widely recognised. In any case, I am not fond of semantic discussions. Nevertheless, it would be sensible to be a little more precise because the concept of politics, especially in political theory, is fairly wide-ranging. For instance, politics is often identified with power: 'it's all politics'; there is the political component of organisations; 'the personal is political'; politics as 'the microphysics of power' (Foucault). I employ a narrower concept of politics. When I talk about politics, I mean the political system. Essentially that means the state, its normative principles, its strategic positions, its operational practices and the different organisations and actors who are either regarded as part of the state or are competing for positions in the state. In other words, political managers, popular representatives, political parties, political ideologies; all the words and things, actors and organisations, principles and practices which in daily use are normally understood by the term 'politics'.

Fragmentation of the Pyramid

Patterns of social meaning, behaviour and organisation are fragmenting. We saw that in earlier chapters and analysed it in terms of postmodernisation. An important consequence has been the decentring of subjectivity at the individual and organisational level. That affects the position of politics because it perceives itself, and is positioned, as the legitimate centre of public decision-making, which is what is meant by political primacy. It is politics that in a finalising sense codifies public decisions, and in an initiating sense modifies social developments. Politics also legitimises public tasks and powers, and distributes them. All political movements of any consequence are in agreement on this, however modestly they define the tasks of politics and the extent of the state. Of course the confessional parties, and especially the theocratic ones, see the primacy of politics as temporary and derivative because they postulate a metaphysical 'centre' beyond this world.

But as the pyramidal ordering of this world fragments, politics, and the primacy of politics in particular, are affected at the level of meaning, knowing, behaving and organising. For even in a pluralistic conception of society, which is fairly widely held, the idea that politics is the final legitimising core of public decision-making has remained intact. Bolkestein (1992) who defends liberalism as a minimalist conception of the state and sees reality as multicentric, remains convinced that the primacy of politics is necessary on principle. But it is precisely this position which has become untenable. In the modernist arguments of autopoietic system-theory, the primacy is socialised through the functional differentiation of subsystems, each of which have specific structures of relevance from which patterns of 'Exclusion und Inklusion' develop. Social unity and cohesion can then only be defined as 'subsystem-specific'.

> Für jedes Teilsystem gibt es dann zwar eine Einheit der Gesellschaft, aber gerade deshalb keine übergreifende, für alle Teile verbindliche Einheit des Ganzen. (Willke, 1992: 43)

The strongly functionalist orientation of system-theory, however, does suggest the possibility of coherence, namely that of functional rationality. From this could be concluded – and often is – that it is up to politics to guarantee the conditions for the maximal development of this functional rationality, and to defend the autonomy of subsystems so that they can form their own intrinsic substantive rationality. But fragmentation in the postmodern sense has followed a different route. The process of fragmentation and the significance of the fragments, understood in the sense of accidental connections, can no longer be contained by the coherence of functional differentiation. The idea of an origin and of history as a linear process, is itself affected by the emergence of fragmentation.

That leads to an existential crisis for politics because it always has to regard fragmentation as a problem and a dysfunction to be combated. Signification, the creation of meanings, in society must never be fragmented because it would mean having to accept a lack of direction. Furthermore, there could be no question of legitimising specific codes of meaning or processes of signification.

So it is the central position of politics which is being undermined by postmodernisation. The organisational aspect is affected because the fragmentation of the administrative system, which can occur spontaneously as well as from restructuring in the interests of flexible specialisation, makes it impossible to control in a co-ordinated fashion. And where frantic efforts are nevertheless made to restore control under the flag of political primacy, it is the developments in the area of ICT (virtualising, horizontalising, deterritorialising) which give these efforts an air of helplessness and hopelessness.

This fragmentation has been recognised by many different observers, though they often describe it differently. Hupe acknowledges the empirical decline of political primacy:

> Partly intended, partly unnoticed, new and possibly temporary equilibria are emerging within the relationships that express the primacy of politics, which are not all necessarily helpful to that primacy. (Hupe, 1995: 65)

Bovens *et al.* (1995) argue that politics has shifted into numerous social domains and developments. Stuurman (1985) characterises the state as labyrinthine because of social fragmentation and subsequent politico-administrative responses to it. Huyse (1994) observes a multiplicity of developments which 'pass politics by'. Guéhenno (1994) points to the disappearance of the pyramid and its replacement by a multidimensional, horizontal network which, incidentally, is itself far from fixed. Whereas some writers still seek to reform politics, to re-orientate politics and democracy, Guéhenno is more forthright in his conclusion:

> This era is nothing more than a way of functioning and we must accept it. Therein lies both the vulnerability and the greatness of the age. There is no political formula that we can employ to ward off the dangers of the postpolitical era. (Guéhenno, 1994: 125)

Fragmentation is, in other words, unavoidable. But in drawing that conclusion one should also be aware of its positive quality for politics. That quality lies in the fact that fragmentation releases politics from its self-appointed obligation to tell a grand narrative. Romanticism, idealism and normativity can then be avoided and politics need no longer pose a threat.

The Tragedy of Unintended Consequences

In spite of the relative decline in, and the problems facing political primacy there is absolutely no indication that the state is in collapse or that the public domain is visibly shrinking. Every citizen is confronted with a vast range of administrative activities, regulations and policies, plans and proposals. But as we have seen, the pattern is fragmented and contingent in its outcomes. Once again the contingency of the outcomes creates problems for politics. They can be considered from various aspects.

In the first place, there is the phenomenon of unintended consequences. They are always unintended from the perspective of the political objectives on which administrative actions are based. Because society is unknowable and the consequences of any administrative action are unpredictable, it is the exception rather than the rule that political goals are ever effectively defined. Formulating them in a more global, more generalised or more abstract fashion, such as defining core tasks whereby politics is implemented through core departments, does not offer a solution. It only leads to many more politically unintended consequences of administrative action.

Secondly, the ambitions embodied in political objectives lead in almost every social domain to calculation, fraud and evasion. A large part of the social reaction to political objectives and administrative attempts to implement them could be characterised as intelligent masquerade: on the surface, political objectives are implemented through all kinds of behavioural patterns and constructions; but in terms of content, existing preferences continue to prevail. This is often done by creating complicated bureaucratic façades designed to remove the content from political control.

Thirdly, defining political objectives becomes increasingly trivial in the light of the complexity and contingency of social developments. On the one hand it means that politics becomes more abstract *vis-à-vis* developing codes of meaning within society. On the other, it leads to an excessive concern with detail such as the impassioned debates in the Second Chamber about tunnels and bridges. Politics is becoming meaningless, from a social perspective inevitably, because its traditional role as the decision-making centre is disappearing.

Fourthly, laying down political objectives is arbitrary and therefore dangerous. It always involves reducing variety and thus implies a strategy of exclusion in respect of specific codes of meaning. Not only does that destroy the self-steering capacities of social domains, it also moralises politics. To define goals is, in its consequences, to establish norms. Goals are not only chosen; they have to be formulated as legitimate or morally correct because otherwise the legitimacy of establishing political objectives is eroded. My earlier objections to normative political theory also apply here.

In the fifth place, social fragmentation leads to a confirmation of the theoretical notion that policy objectives only become real in their implementation. Policy implementation is in fact policy-making. It is highly exceptional if the implementation of a policy actually achieves the *a priori* objectives of that policy. And if that exception does occur, it is usually by accident. The strategic significance of policy implementation (see Huigen, 1994) therefore also contributes to unintended consequences.

In the sixth place, unintended consequences lead via a wide range of feedback mechanisms to the formulation of new objectives: the 'law of policy accumulation' (In 't Veld, 1984: 103). However, the conditions which give rise to these unintended consequences can not be resisted because they arise from the fundamental contingency of social processes. Adjustment, fine-tuning, central co-ordination, all are tragic attempts to pull politics out of the mire of a society which lacks precedent. They are tragic because they only increase fragmentation and its resulting contingencies. They are problematic because they make politics pathological and can thereby distort the connections between politics and society.

The Advantage of Indifference

Confronted by fragmentation and contingency, politics can only be indifferent to the outcome of processes of signification, of meaning-creation in which administrative actors are also involved. Socially generated meanings are accidental connections which can neither be defined *a priori* nor be predicted. Political objectives have thus become pointless because coherent meanings for a decentred reality can not be formulated from a centre. And they are dangerous in so far as other specific meanings have to be forbidden or marginalised. This process of exclusion always affects potentially intelligent and attractive connections which are then forced into becoming subversive. The inevitable consequence of all this is that politics becomes empty of content, that it no longer formulates any preferences of substance in respect of the fragmented social domains, that it becomes 'unprincipled in principle' and 'neutral', as De Wit (1995) called it in a reaction to Ankersmit and Schmitt. Depoliticisation is an inherent tendency in democracy in so far as it is perceived as a mechanism for reconciliation and negotiation:

> The basic idea here [in the work of Schmitt and Ankersmit] is that an effective neutralising of serious political disagreement and conflict is only possible on the basis of a third party, a neutral area. It is only 'neutral' in the sense that it is *above discussion* and serves as the undisputed basis for consultation and discussion. And to that should be added: it is neutral only *so long* as it is removed from discussion, competition and political conflict. (De Wit, 1995: 161)

According to those two writers, neutralising politics in its classical, ideological shape is an historically inevitable outcome of the modernising process. According to Ankersmit, the neutral area is the state and it must 'take the lead' (Ankersmit, 1994a). Precisely because it is neutral, only the state is in a position to resolve social problems. The difficulty with this conclusion naturally is that problem-solving implies some objective, however functional. And objectives conflict with neutrality, unless we assume some minimal consensus on a limited number of goals. But even that assumption is problematic because of the process of fragmentation and the contingency of unintended consequences.

Political indifference is not only inevitable if the trivial, arbitrary and harmful effects of setting political objectives are to be avoided, but also desirable for the enrichment of fragmentation and contingency so that a maximum variety of connections can be created. Political indifference is, moreover, necessary in a decentred reality. For if social developments are no longer determined from one centre, or even several centres, there can be no central criteria by which to judge those developments. Passing such judgement can no longer fall to any specific institution definable as a place. Nor can it fall to different institutions to which politics may have shifted. For it is not a shift which results in a scattered, but definable, location of politics as Bovens *et al.* (1995) seem to suggest, but a completely diffuse displacement which can not be traced, can not be controlled, and which is always unexpected and surprising (and not always pleasantly). Its diffuse character ensures that politics, in its totality of actions, processes and actors, has to be indifferent. It must be a politics without properties and display a wide variety of spaces which are empty of power.

10.3 THE APOLITICAL PUBLIC DOMAIN

Politics without properties implies a public domain that is apolitical, which is to say that it can be conceived of as a domain that does not embody any preferences of substance. The content is continually emerging contingently within processes of social signification which are decentred and fragmented. The public domain then can not be seen as a space which is nameable or recognisable; it has rather the character of a fabric or, as we observed earlier, a rhizome. It has no beginning or end, it has no linearity, and it has no centre. Furthermore, the fabric is untidy: it has no clear structure nor a coherent style. It takes the form, primarily, of infrastructures for information, communication and decision-making. The apolitical public domain is not even consistent because it is pluralistic. And it has no intention of producing consistency. The apolitical public domain has no territory because societal processes of signification have been deterritorialised.

> The deterritorialization of capital, information, opinions and war have combined to undermine the very notion of sovereignty: a monopoly of public power within defined territories, able to negotiate with other similar monopolies. (Mulgan, 1994: 203)

Finally, the apolitical domain has no ideological foundation and does not attempt to formulate one. This public domain can never be monopolised, (either empirically or normatively) by a bounded rationality.

The Empty Place of Power 2

Earlier in Chapter 5 I explored the theme of the empty place of power. This terminology is used by Foqué (1992; Kuypers, Foqué & Frissen, 1993) to make clear that the politically legitimised rule of law should on principle be thought of as empty:

> It is a place which no longer belongs to anyone and may not be monopolised by any specific ideology or philosophy. The identity of a society, in a democracy, is in principle intangible and latent (…) (Foqué, 1992: 29)

In a discussion with Foqué I suggested that the political reorientation towards the procedures and structures of decision-making and the associated substantive decline in political primacy reveals an astonishing kinship with this idea of the empty place of power. Meta-steering could then mean, on the one hand, creating and encouraging conditions in which one can talk and negotiate about power, or on the other, keeping the space of power empty because meta-steering rests on indifference to content. (See Kuypers *et al.*, 1993: 55ff.)

Contrary to Foqué, I consider it much more problematic that the empty place of power is transcendent and that its transcendence extends to the establishment of the general interest. A general interest, moreover, that is primarily conceived of as a potential and necessity of principle, as a 'symbolic externality' which never coincides with the concrete actuality of the political facts of an ideology (Foqué, 1992: 30–31). But how can we imagine that transcendence without a metaphysical principle or an ideologically grounded criterion? Neither is the general interest a self-evident, pre-defined or neutral concept, unless we understand it as contingent and immanent in social and historical developments.

Foqué emphasises this transcendence to escape being accused of 'postmodern opportunism' and to contest the notion that the state can be broken up (Foqué, 1992: 30)

In my view, processes of social fragmentation give rise to an expansion and pluralising of the space constituting the place of power. I see the expansion primarily as a diffusion of the public domain so that no spatial connotation remains. Pluralising implies that there is no longer any question of the public domain being conceived of as coherent or as a centre. The public domain is no longer hierarchical, either in its distribution of positions, powers and responsibilities, or in the sense of a hierarchy of ends and means. In the idea of the general interest, 'general' is also pluralised and can only ever be the temporary result of social processes of signification. But the pluralising does mean that the temporary connections of social codes will never be fixed and are always renegotiable. At the same time, the public domain is neutral in respect of the substantive outcomes of those negotiations because it does not embody a criterion, not even that of the public interest, by which those outcomes can be judged normatively. The only possible judgement left to politics is whether the public domain is able to honour fragmentation and contingency. And that judgement is in principle unprincipled.

The public domain, therefore, can not be seen as integrated or as a coherent pattern. Precisely because it is immanent in the social processes of signification, it is fragmented and contributes to contingency and complexity rather than being able to combat and control them. Knowing mainly produces a state of not-knowing. Information and communication in the public domain can admittedly lead to connections that are either spontaneous or created in a process of decision-making. But the spontaneity is inevitably followed by new and different forms of spontaneity, while the decision-making always takes on the characteristics of negotiation whose results are always renegotiable. The public domain is, to echo Rorty, an infinite collection of vocabularies and not a politically final vocabulary.

Information, Communication, Decision-Making: Infrastructures

I would describe the apolitical public domain as the infrastructures for information, communication and decision-making which are present and developing within society. They are the administrative structures which arise through processes of administrative restructuring and policy reorientation. They also include associated ICT infrastructures. These infrastructures constitute a public domain in which politics has become displaced or diffuse. Being infrastructures, they are conditional and politically neutral. They do not determine any specific outcomes of substance unless politics intervenes with strategies of inclusion and exclusion. Van Gunsteren and Andeweg (1994: 124) rightly identify this as a reason not to resist the decline of political primacy. According to them, the core value of democratic politics is 'to avoid dictatorship and to establish the republic' (Van Gunsteren & Andeweg, 1994: 108). Avoiding dictatorships has echoes of Rorty's idea of avoiding cruelty on the one hand, and keeping the space of power empty on the other. Establishing the republic might indicate the creation of infrastructures for information, communication and decision-making as we have described them. Van Gunsteren and Andeweg will probably object to the neutral or apolitical significance which I attach to the public domain. They will probably still want to link the republic with the realising of values and objectives. If these values and goals are formulated so modestly that they merely concern the conditions for pluralistic decision-making, information and communication about values and goals in fragmented domains, then there are only differences of nuance between us. But the infrastructures which I characterise as the public domain can not be conceived of as a basis for decisionism. The illusion that our decisions determine the course of history – however useful that may be for codes of meaning – can not be the principle for defining politics. Politics ought to be ironic in the face of that illusion, it ought to judge infrastructures by the extent to which they do justice to plurality and thereby produce fragmentation and contingency. This means that political attention shifts from the message to the medium, to paraphrase McLuhan's adage. The media for social signification, in particular those which can produce public meanings, are application points for politics.

Politics, then, has no primacy in legitimising specific public meanings, but helps to keep the media open and plural in the recognition that the media themselves make up part of the public domain. Openness and plurality assume, however, that these media are neutral and apolitical, and therefore non-selective and indifferent in respect of the message.

Anti-Utopia

This narrative is not utopian. Its motif is not the possibility of a better world. And that lies at the heart of the idea that the infrastructures constituting the public domain are neutral and apolitical. The narrative is not utopian because on the one hand fragmentation and contingency are unavoidable, and on the other the consequence of pursuing a normative grounding through politics is more dangerous than the undoubted sincerity of the aspiration. The orientation toward infrastructures which I advocate for politics – not as an exclusive task and acknowledging that the infrastructures are merely the public domain and not politics itself – differs from Salet's advocacy of a politics which renounces the one-sided rationality of objectives and the production of instrumental policies which it leads to. He argues that politics should be primarily concerned with 'a robust framework of principled legal relationships' and policy domains with 'establishing and maintaining fairly durable normative patterns' (Salet, 1994: 112). By that Salet is referring to an institutional infrastructure. But he emphasises the responsibility of politics for its normative content. I reject that normative orientation of politics because normativity is always the outcome of social codes of meaning and should not be *a priori*. *A priori* normativity runs up against our earlier objections to exclusion, principled selectivity, and the neglect of potential intelligence and creativity. That is why this narrative is anti-utopian where it concerns politics: 'We infallibly sense that everything which is presented as *non-negotiable* can threaten us and is potentially hostile' (De Wit, 1995: 163). If we define politics as attention for public infrastructures and as the attempt to condition their openness and plurality, then nothing can be non-negotiable in politics. We might as well abandon our dreams of a better world:

> However, unlike a legal machine, this community of values or rather a mosaic of overlapping communities, is, like all democracies, far from rational. It is capable of combining irrationality, prejudice, unethical behaviour and unprovoked aggression. (Mulgan, 1994: 194)

But the possibility of unpleasant outcomes should not seduce us into making politics normative or utopian, because then the outcomes will be predictably unpleasant.

10.4 POSTMODERN POLITICS

But is postmodern politics conceivable? Even conceptually we run up against a contradiction between what we might understand by postmodern or postmodernistic and the many obvious meanings which the term 'politics' seems to have. At the same time, there is the history of politics in our type of society which has led to the particular manifestation of the welfare state. By association it is immediately assumed that politics is tied up with power over the collective, with ideological legitimisation, and with the extensive machinery of bureaucracy. Politics then comes into conflict with what I have described as the characteristics of postmodernisation: fragmentation, contingency, decentring, virtualisation. Can that contradiction be bridged? Is it desirable or even possible to bridge it? In itself, politics may as well disappear – no historical phenomenon has a right to eternity. But we should distinguish between the historical manifestation of a concrete political system and the behavioural practices which it institutionalises. I hesitate to speak of 'basic anthropological patterns' but even without adopting an essentialist standpoint we can assume that the public domain – information, communication, decision-making, conflict in public affairs – will long outlive the concrete form and significance of any political system.

We must then see whether, in the light of postmodernisation, politics can take on another form and significance which will do more justice to the social developments we have been describing in this book. I have argued for the necessity of this by pointing out the trivial, arbitrary and harmful nature of a politics which persists in its traditional forms and attitudes. Of course, those who see these social developments as themselves trivial, arbitrary and harmful – an obvious conclusion from a modernist or normative point of view – will prefer to look for a new or renewed moral foundation for politics.

Here another course has been chosen. On the one hand, I join Willke in his search for the possibility of a politics which does justice to contingency:

> Eine anders Auswirkung der Unerschöpflichkeit von Alternativen zu den Alternativen ist die Notwendigkeit einer sekundären Stabilisierung von Politik, die allerdings so geschaffen sein muß, daß sie nur auf eine bestimmte Dauer bzw. auf Widerruf stabilisiert, denn sonst würde sich wieder nur eine einzig mögliche Ordnung festfahren. Es geht also um eine Form der *Kontrolle von Kontingenz*, welche die paradoxe Leistung einer Stabilisierung der Variabilität und einer Wiederauflösung des Stabilen zugleich erbringt. (Willke, 1992: 37)

On the other hand, I want to show that Rorty is mistaken in his argument that irony in public affairs is impossible. Postmodern politics must seek out and make possible just this kind of public irony. Willke, too, sees that as the solution to the paradox to which he refers.

A Grammar for Decision-Making

Previously (in Kuypers, Foqué & Frissen, 1993) I have suggested that in the light of social fragmentation and contingency, and the consequent necessity for indifference to substance, politics should formulate a grammar for public decision-making. This grammar should relate primarily to creating norms for social decision-making arrangements and grounding them in what Foqué calls transcendence. Politics would then be the institution that maintains the anarchistic chaos of social variety and is responsible only for the progress and quality of decision-making (Kuypers, Foqué & Frissen, 1993: 56).

However, that opinion ought to be modified: it is still normative and assumes, moreover, a hierarchical position for politics. I should now like to choose a different point of departure. That point of departure is the importance of style, which we observed earlier. In the public domain too, style is the aspect of ideology which is becoming increasingly dominant over the normative aspect:

> Nowadays, the citizen judges the government less by ideological criteria, and is more aware of the 'style' and the 'form' of the state's actions (De Wit, 1995: 158).

Tops (1994; 1995) in his study of politicians and administrators in local government also shows that style is becoming more important and that it may well be the source of a new political legitimacy. Style has an important bearing on the design of social decision-making processes.

Postmodern politics can be concretised at two levels. Firstly in the insight that a wide variety of decision-making styles exists in the public domain. In other words, there are many grammars for decision-making which can compete with each other, which frequently change, and which can sometimes make connections with each other. Here it is evident that the judgement whether a connection has been made and what its quality is will vary according to the decision-making style. Secondly, politics must validate its own operation by maintaining this variety of decision-making styles. It must defend the conditions which make maximal plurality possible. In defending plurality there are again different possible styles and it is about those differences that political competition could be organised. Such competition would be much more aesthetic than ideological.

Postmodern politics is largely a question of form, of externals. By that I am not referring to the widely detested personalising, Americanising or 'media-ising' of politics, though there is little wrong in that, but much more to the application points of political interference: they remain on the outside of substantive codes of meaning in fragmented social reality (the edges), and on its 'underside', the infrastructures. The position of politics is thus equal or subordinate, rather than hierarchically superior. It is indifferent to substance, which is empirically unavoidable and normatively desirable, and it tries to protect and enlarge the potential for the social development of grammars for public decision-making. That is the basis of Willke's notion of 'Kontrolle von Kontingenz': a control that is not aimed at reducing and dominating contingency, but at making it possible. That is exactly why politics has to take up an ironic position. Only from a tragic point of view, do fragmentation and contingency become problematic.

> Für den ironiker dagegen eröffnet die unabänderliche Distanz zwischen den Systemen den Spielraum für die Möglichkeit einer Akkordierung von Kontingenzen, wenn erst einmal klar ist, daß dies nicht aus der Position einer höheren oder überlegenden Rationalität (welchen Beobachters oder Akteurs auch immer) bewirkt werden kann, sondern allein aus der Spiegelung (Spekulation, Reflexion) der äußeren Distanz in einer *inneren Distanz der Systeme zu sich selbst*, die ihnen die eigene ironische Position eines Spielens mit ihren Kontingenzen erlaubt. (Willke, 1993: 321–322)

If we ignore the functionalist orientation which rouses the suspicion that even for Willke there exists a superior rationality, namely that of modernist functional differentiation, we can see immediately the connection between styles of decision-making and grammars on the one hand, and irony on the other. For politics can only do justice to the variety of decision-making styles if it does not choose between them, if it adopts an ironic position in respect of the qualities and styles of decision-making and of the outcomes which they generate. Furthermore, it must be ironic in respect of alleged connections between specific styles and specific outcomes. The quality of the styles lies precisely in the contingency of their outcomes. Only in honouring the variety of styles can that contingency be realised. In another respects, politics joins up with what I said earlier about a postmodern public administration which conceives of technocracy as art. Perhaps Tromp will be proved right in his observation that in the Netherlands, politics is thought of primarily as administration.

The Aesthetics of Technocracy

In the modern politico-administrative system bureaucracy counts as an instrument of politics. Bureaucracy embodies the control technology of political power. But the political primacy over bureaucracy expressed in this subject–object relationship has been undermined by its inherent tendencies as well as by administrative developments, professionalisation and technological developments. These have been explored extensively in previous chapters. Postmodern politics then has significance for the special relationship between politics and bureaucracy. Three aspects must be considered:

- The bureaucracy (or better, bureaucracies) is a constituent part of the public infrastructure and thus deeply interwoven with the fragmented processes of social signification. Previously described administrative developments are an expression of this and they are further driven by technological developments in the direction of horizontalisation, autonomisation, deterritorialisation and virtualisation. There can be no question of an instrumental relationship in the sense of a political master and a bureaucratic servant.
- Bureaucratic knowledge is increasingly becoming technocratic knowledge through professionalisation and technology. But that technocratic knowledge differs from the rationalistic principles of technocracy which Fischer criticises (1990). It is rather multi-rational and attempts to understand and honour contingency. Hence the growing bureaucratic orientation toward network approaches to steering and away from involvement in substantive policy.
- For this reason among others, technocratic knowledge is also ironic. It questions the possibility of grand narratives and final answers. The distinguishing features of steering arrangements must be openness and variety so that plurality will be honoured and potential outcomes, in the sense of connections, will not be totalitarian.

Technocracy, through being interwoven with social domains, has become significantly autonomous and fragmented. It is a form of art because it creates conditions for social codes of meaning and does so in an ironic style.

Postmodern politics will have to recognise the aesthetic character of bureaucracy. And where politics in general approaches public infrastructures from the point of view of contingency and irony, it will have to do the same in respect of bureaucracy, which is, after all, a part of those infrastructures. This has a number of consequences.

In the first place, politics will have to acknowledge that the public infrastructures, of which bureaucracy is a part, have become highly self-regulating and self-steering through processes often interpreted as 'a displacement of politics' (Bovens *et al.*, 1994 and 1995; Depla & Monasch, 1994).

Secondly, in a postmodern sense bureaucracy already fulfils a political role in those public infrastructures, or would be able to do so. Bureaucracy, in the sense of technocratic knowledge, is increasingly fulfilling the task of organising and designing processes of signification. In this it tries to respect a variety of decision-making styles and itself employs changing repertoires.

Trusting in social autonomy then often implies withdrawing from (institutional) politics.

In the third place, postmodern politics, in so far as it wants and is able to play a role in respect of the bureaucracy, for instance in the organisation of patterns of responsibility, must perceive this role as a form of orientation. It must orientate the bureaucracy toward the desired openness and variety of public infrastructures. It then has a particular responsibility to insist on that openness and variety just because contingency and irony are desirable. Once again, such responsibility is aesthetic in nature because forms and styles are being judged, not outcomes. In practice it means that postmodern politics leads to a significant modification in the position of the traditional political system, particularly in so far as that position is believed to be a centre. Postmodern politics expresses the fact that politics is dispersed in a diffuse manner throughout many public infrastructures. In so far as politics as an institution any longer has a role, it is aesthetic in nature since it focuses on the forms and styles of public decision-making processes. Information and communication and their associated technologies play a prominent role in this.

Information and Communication

Public infrastructures have always formed patterns of information and communication. Historically they have been increasingly influenced by technologies. So it is surprising that today's political system has changed so little since the time when the telephone had yet to be discovered. It means, as I have so often argued in this narrative, that massive incongruences have developed between traditional politics and public infrastructures, with all their administrative and technological components.

Public infrastructures have changed as a result of social developments, administrative patterns of response and the characteristics of its technologies. I have described the change as tendencies to horizontalisation, autonomisation, deterritorialisation and virtualisation. Postmodernising is a process in which information and communication technologies have a real role to play and the importance of this is finally starting to get through to traditional politics. (See Ministerie van Binnenlandse Zaken, 1995 and Ministerie van Economische Zaken, 1994.) But an instrumental orientation still predominates and ICT is still seen as a new tool that is good for the economy, useful for administration and handy for politics. It is recognised that ICT tends to disturb existing institutions, but for some reason politics is assumed to be an exception.

Here the position is seen differently. ICT and the infrastructures developing along with it are not primarily tools but are the new public infrastructure. Social signification is tied up with this and will take shape increasingly within these ICT infrastructures. Given the explosion of signification taking place, fragmentation and contingency will increase rather than decrease. We can be sure in advance that political domination is illusory. Postmodern politics disappears as it were in the infrastructures. The technoculture of cyberspace is also a political culture so a retreat from institutional politics is inevitable. But that insight does not imply that no political role will remain. Encouraging the openness and variety of electronic structures is a political interest. Plurality and contingencies must be made possible in a conditional sense. That will incidentally sometimes require a political retreat via liberalising and deregulation as well as the prevention of monopolies and exclusivity in signification. The grand narrative of politics may of course be replaced by another grand narrative.

For this reason too, irony is called for: infrastructures for information and communication need to be stimulated, but also to be judged according to whether they produce exclusion, reduce variety, and attempt to control contingency and obstruct fragmentation. That could be the value-free normativity or the principled lack of principle of postmodern politics. That insight is the core of political postmodernism.

10.5 POLITICAL POSTMODERNISM

In the preceding sections I have described, partly theoretically and partly speculatively, the consequences of the process of postmodernisation for the form and content of postmodern politics. The perspective adopted is virtually that of contingency theory, namely that the political system can only establish meaningful connections with the social domains of signification if it respects those domains (fragmentation, autonomy, and contingency). Furthermore, I have advocated postmodern politics as an alternative to a politics inspired by normative political theory. To avoid any accusation of the functionalism with which I have reproached system-theory, it would therefore be sensible to sketch the possibility of a non-normative political theory which can, as it were, formulate and perhaps legitimise the self-perception of postmodern politics. Such a political theory I shall name political postmodernism.

Small Narratives

In opposition to the universalist claims of normative politics, postmodernism sets up a 'politics of small narratives' (Mascarpone, 1992). Universalism is contested not only because postmodernism expresses theoretically that the belief in grand narratives has come to an end but also because philosophically it rejects any grand narrative. Grand narratives, or meta-narratives, are metaphysical in character and believe in essentials which politics is supposed to convert from dream to reality. The dream is metaphysical because its image of the future is not merely an accidental one, it is the only possible and just one. That metaphysical conviction turns history into the realisation of a teleological principle. In modernist political theory, for which Habermas is an important spokesman, its teleology lies in justice and reasonableness. The Enlightenment, however repugnant in its historical contingencies, holds out the promise of domination over those contingencies. The inter-subjectivity of power-free communication is the condition under which the subjectivity of society can be realised.

The narratives of postmodernism are much smaller. They start out from the heterogeneity and pluralism of innumerable language games which are irreducible to each other and are outside any hierarchy of integrating criteria. That applies even more urgently to political narratives which have an intrinsic tendency to be 'grand' in the sense of wanting to be totalising. Political narratives are, after all, about the good society and about the duty of politics to bring about the good society or at least to embody the striving to do so. That is why political narratives are so often utopian. The small narratives of political postmodernism put into words the disbelief in the modernistic ethos of domination and justice.

> A postmodernist politics and political culture would highlight a dissatisfaction with modern politics, its sameness, customary allegiances, its predictability, bureaucracy, discipline, authority and mechanical operation, and would stress the emergence of a politics featuring difference, dealignment and realignment, unpredictability, freedom, delegitimization and distrust, power and spontaneity. (Gibbins, 1989: 15-16)

But not only can political postmodernism be seen as a collection of small narratives criticising the grand narrative, its very conception of politics sees the function of politics in postmodern society as making small narratives possible. That is to say that politics no longer seeks homogeneity from its position as the public decision-making centre, but rather follows and stimulates heterogeneity and attempts to prevent the hegemony of any kind of specific political content. That is not the same as the modern democratic belief that minorities deserve protection and respect, but much rather expresses the idea that postmodern societies consist *only* of minorities. Political postmodernism accepts fragmentation and tries to make spontaneous connections possible not primarily by organising them but by removing any obstacles to them.

Political postmodernism propagates modesty of political pretensions and ambitions. Modest pretensions because reality has become too complex and fragmented to control centrally and politics is spread too diffusely. Modest ambitions because neither politics or any other institution can be allowed to ground the legitimacy of one narrative at the expense of many others. The legitimacy of narratives is determined in the various subcultures; morality is a local matter. One task of politics is to prevent the definition of subcultural legitimacy in fragmented domains from assuming the illegitimacy of other subcultures, and the assumption that the political system is the arena in which to resolve the conflict. The narcissism of individual right must be kept individual with the help of public pragmatism (Rorty, 1991a: 210). It is in that sense that political postmodernism is also apolitical because it refuses to position itself at the level of a political system or discourse that can resolve political differences of opinion (see also Van Reijen, 1988: 219-220). If postmodern politics wants to do justice to plurality and contingencies, it will above all have to be ironic. And irony is apolitical.

Public Irony

Political postmodernism is ironic not only because it rejects the possibility of grounding politics in a grand narrative, but also because it wants to make of that rejection a design-principle of politics and public infrastructures. The anti-metaphysical standpoint of postmodernism can not mean that a rigorous anti-metaphysics is set against the rigour of metaphysics. Its anti-metaphysics is ironic, on the one hand because it is superficial in the sense of not wanting to be fundamentalist, and on the other hand because it tries to respect the historicity and provisional nature of every viewpoint. Notwithstanding Rorty's disbelief in the possibility of public irony, I regard it as a necessity for politics to be ironic. For if politics becomes serious, there is always the temptation to be idealistic and to regard politics as the pursuit of that ideal. Politics would then have to oppose fragmentation and to try to control contingency by adopting some substantive guideline as a basic principle for its actions.

The irony of political postmodernism therefore relates in the first place to the truth and justice of political narratives. In political postmodernism truth is deconstructed by tracking down distinctions, pointing out the connection with power, and emphasising the local nature of knowledge. Justice in political postmodernism is a moral sentiment which may be valid for specific lifestyles and individual citizens but over which politics has no need to reach agreement in order to act. A liberal society can be satisfied with its liberal behaviour without having to ground it in morality. Put more strongly, to give liberalism a moral grounding would conflict with liberalism itself because once again contingencies would be excluded and the autonomy, subjectivity and self-referentiality of social domains would be damaged. (See Willke,

1992: 323) That is why Rorty's principle of avoiding cruelty as the goal of liberalism is not disappointing but politically powerful. All kinds of attempts to tie postmodernism to the moral validity of a universal principle of justice, however pluralistically formulated, have foundered in the past. (See *e.g.* Boutellier, 1994; Heller & Fehér, 1988; White, 1991.)

In the second place, the irony of political postmodernism relates to public infrastructures and the patterns of information, communication and decision-making that they make possible. The irony consists in recognising the styles which regulate different practices of signification. Politics has to recognise them as styles, which means that they are not reducible to each other and that no criterion exists on which to base a preference for any politically specific style. Neither is there any aesthetic criterion since the notion of *avant garde* has disappeared. Furthermore, politics must recognise the temporary and provisional nature of all styles, and work for a design of public infrastructure that remains open to new styles and connections.

Thirdly, the public irony of political postmodernism relates to the position of politics itself in postmodern society. That position is no longer a centre, but is found underneath or at the edges of the domains of social signification. This implies a modest and reticent position for politics. Irony requires that reticence and modesty should lead to minimalist politics – contrary to what some may think, much of social signification is not political, – and to apolitical politics. To make politics possible, politics must often be apolitical. Only in this way is it possible to avoid the dangerous idealism and totalitarian normativity which we saw in the discussion of normative politics.

Hedonistic Pluralism

Political postmodernism has a close affinity with ideas of pluralistic democracy and anarchism. (See also Mascarpone, 1992: 99ff.) But it differs from pluralistic democracy because it desires to change the institutional structure of the political system, which pluralism sees as hierarchical and grounded in the belief that democracy is morally superior. It differs from anarchism because it has no illusions about a better society of free and equal citizens. It is not utopian; indeed, in its rejection of grand narratives it is anti-utopian. Political postmodernism is, however, pluralistic but then in the sense of recognising fragmentation and contingency. The search for connections is no longer a political goal in the sense of creating consensus as an input for politics; it is much more a question of making connections possible as accidental results of a fragmented process. And connections can only be made possible if plurality is maintained, thus making every connection continuously renegotiable. The pluralism which political postmodernism champions is therefore contingent or, in other words, chaotic and anarchistic.

The anarchistic character of political postmodernism rests in the first place on its diagnosis of society. It regards society as the constantly contingent outcome of processes of fragmentation. This fragmentation moreover leads to decentring in every domain so that it is no longer possible to speak of locating power as an oppressive institution. That does not mean to say that power vanishes but that it too is fragmented.

In the second place, its anarchism lies in the desire to make political institutions more congruent with the characteristics of postmodern society which, being fragmented and contingent, demands a politics that is anarchistic, non-hierarchical, without congealed power relationships, and preferably libertarian in orientation.

In the third place, political postmodernism is anarchistic because it rejects the subordination of actors and organisations to political ideals. It confines itself, at least in my interpretation, to attempting modestly 'to avoid cruelty'.

Hence my emphasis on aesthetics and irony. And in that sense, the pluralism or anarchism of political postmodernism is hedonistic. The 'highest good' which it strives for is what is pleasing, which in concrete politics boils down to avoiding exclusion, aggression, needless conflict, and encouraging creativity, autonomy and the capacity to learn. For that a public infrastructure for information, communication and decision-making is essential. Because it contains the conditions needed to stimulate creativity, autonomy and the capacity to learn, it is pre-eminently suited to realising pleasure. Irony is an essential political attitude because we can only actualise what is pleasing if we are relativistic in respect of all moral claims which pursue political realisation. Such relativism is not a morally superior attitude, as Rorty rightly points out (1991b: 202) but the only way to avoid the danger of imposing moral superiority. And it is especially politics that carries that danger if it pursues what is just instead of what is pleasant, and prefers truth to aesthetics.

10.6 DIGITAL AMBIGUITIES 4: THE VIRTUAL STATE

This narrative is nearing its end. An author often considers that to be a good time to draw conclusions. I shall not do so because I have not been able to tell a rounded narrative which leads to clear and evident conclusions. There are without any doubt loose ends, speculations, presumptions and unfounded assertions. I shall therefore devote these final pages to the image of the virtual state. An image brimful of ambiguities and uncertainties. But it is nevertheless an image which corresponds to the developments that I have been describing. Developments in politics, governance and technology about which I wanted to tell a postmodern narrative.

In public administration we can see two patterns simultaneously: one of modernisation and one of postmodernisation. The pattern of modernisation flows from the many changes, experiments and innovations introduced in recent years and which have been intended either to remove classic shortcomings of bureaucracy or to refine further the structure and functioning of the administration. I have called them amendments and differentiation of bureaucracy and they involve modernising the administration by rationalising and perfecting the technologies of intervention.

But in the process of modernisation, postmodernisation is already visible; it is there where differentiation becomes hyperdifferentiation and where the limits of monocentric organisation come within sight. (See Crook *et al.*, 1992.) In the words of Beck, Giddens and Lash (1994) it is where modernisation becomes reflexive and effects strike back at their causes and work against them. Modernisation then becomes ambiguous, without clarity, without linear development, and no longer susceptible of conception or control from (the notion of) a centre.

I called that a reversal for bureaucracy: the pyramid fragments and becomes an archipelago; policy processes become circular; contracts and self-regulation replace regulating decrees. Postmodernisation can be understood as a radicalisation of the modernisation process in public administration; but it is a radicalising which leads to a qualitative break.

Information and communication technology reveals the same picture: on the one hand, the technology is the symbol of modern culture; on the other, aspects of it are undermining that culture. ICT embodies the classical modernist values of standardisation, formalisation and specialisation which lead to further refinements and potential for domination. Informatisation is rationalisation and progress; modernist self-perception is most perfectly expressed in its techniques.

In an organisational sense the technology contributes to perfecting the pursuit of domination and the completion of transparency. Culturally, the processes of bureaucratisation and informatisation are closely related and they lead to ever-increasing differentiation. But this differentiation is such that simultaneous processes of horizontalisation and virtualisation occur that make the classical mechanisms of bureaucratic arrangement and hierarchical control obsolete. Organisations and organisational relations become virtual through changes in time and space. And the instrumental position of technology, assumed to be a prerequisite in the modernist pursuit of domination, can no longer be sustained because of its autonomous power which diminishes subjectivity in the sense of anthropocentrism and decisionism.

Both in and through technology, the process of modernisation is radicalised into postmodernisation. But ambiguity remains because both processes continue at the same time and in many respects are the reverse side of the other.

Informatisation in and of the administration is a concrete manifestation of this ambiguity. The *Wahlverwandtschaft*, the elective affinity, between bureaucratic culture and informatisation is visible in the technocratisation of administration to which the expansion of ICT leads. (See also Van de Donk, Snellen & Tops, 1995.) The new image of the bureaucracy is the infocracy: the bureau and the hierarchy have been replaced by the architecture and infrastructure of the technology. The dream of perfect bureaucracy appears to have come true because the limitations inherent in bureaucracy have been removed. Individualised masses and dynamic stability have been actualised.

Change, experiment and innovation in the administration have been made possible by ICT. Controls and services can be made to measure. Ambitions of domination reveal themselves not so much in the repressive shape of the surveillance-state as in the noble intentions of the anticipating state.

But the same changes, experiments and innovations are being pushed further in the direction of horizontalisation, differentiation and autonomisation. And that creates problems for any attempt to dominate from a monocentric mind-frame and an organised bureaucracy. Deterritorialisation and virtualisation produce an archipelago of administrative relationships, a fragmentation of structures and processes, and a decentring of processes of decision-making and signification. The Orwell who flits through Athens has long ago lost the recognisable and stern features of Big Brother, and has donned the ambiguous grin of an intangible many-headed monster.

The virtual state is a tragic problem, particularly for politics. On the one hand, this is the consequence of the political ambition to dominate. Policy and control can be made more finely meshed. But achieving that ambition strengthens bureaucracy and technocracy, and politics is displaced into new arenas. However, this displacement of politics, through technological and administrative developments, is also at the same time a fundamental change of shape. The image of politics as a social decision-centre splinters in the fragmentation of postmodernisation; displacement is one manifestation. There is some political reflection going on about this but the suggested strategies would test the skills even of Baron von Münchhausen. Even if one wanted to pull oneself out of the mire by one's own hair, the head on which the hair is growing has vanished.

Of no greater help is the futile romanticism of communitarism. It is superfluous because the forming of communities, certainly in a virtual form, no longer requires a political centre, and it is dangerous because it tries to combat the loss of political primacy through a strategy of moralisation, forgetting that the loss of primacy in the first place is largely owing to the shortcomings of the grand narrative. Freedom does not need morality any more than sex needs love (Hirsch Ballin: 1995).

Politics will have to accept the virtual state. It is an inevitable consequence of both modernisation and postmodernisation. Furthermore, it corresponds to the nature of social developments. In the economy, in organisations and in culture, we can see a comparable pattern of fragmentation. The modern contract society is an archipelago and its culture is staccato in character. The modern has become postmodern.

The virtual state is a metaphor for the complex and contingent connections between politics, administration and technology in the 'postmodern condition'. Postmodernism as a theoretical orientation attempts to describe these connections and employs a vocabulary which tries not to be totalising or metaphysical. That is why fragmentation, connectivity and contingency were given such a prominent place in my description of politics and administration in cyberspace and my proposals for a postmodern theory of administration and politics.

Connections can only be grasped in a non-reductionist, non-totalising perspective if we understand them to be a product and a form of fragmentation. Empirically, connections are contingent. Administration and politics must be designed to leave enough space for contingency and when necessary to enlarge it. Administrative and political theory are then proposals for intelligence and aesthetics: how politics and administration might cope with unpredictability and uncontrollability in an intelligent and pleasing fashion. To achieve that, one needs irony and the ironic reflection which small narratives have to offer. I have tried to tell just such a narrative.

Bibliography

Abma, T.A. (1994), '(Bestuurs)wetenschap in een plurale, postmoderne samenleving', *Bestuurwetenschappen*, (1994) **5**, 443–445.

Algemene Rekenkamer (1995), *Verslag van de Algemene Rekenkamer over 1994*, Tweede Kamer, vergaderjaar 1994–1995, 24130, nr.3.

Ankersmit, F.R. (1994^a), 'De staat moet weer het voortouw nemen', *NRC Handelsblad*, 1 october 1994, 9.

Ankersmit, F.R. (1994^b), 'Veronderstelt de politieke filosofie de ethiek?', *Justitiële verkenningen*, 20 (1994) **6**, 123–137.

Attali, J. (1992), *Millennium. Naar een nieuwe wereldorde.* Utrecht/Antwerpen: Kosmos.

Baakman, N.A.A. (1990), *Kritiek van het openbaar bestuur*. Besluitvorming over de bouw van ziekenhuizen in Nederland tussen 1960 en 1985, Amsterdam: Thesis Publishers.

Backx, H.A.M. and E.M.H. Hirsch Ballin (eds) (1990), *Recht doen door wetgeving*. Opstellen over wetgevingsvraagstukken, Zwolle: W.E.J. Tjeenk Willink.

Baudrillard, J. (1985), 'The Ecstasy of Communication', in H. Foster (ed.), *Postmodern Culture*, London: Pluto Press, pp. 126–133.

Baudrillard, J. (1988), *Selected Writings*, Edited and Introduced by Mark Foster, Cambridge: Polity Press.

Bauman, Z. (1992), *Intimations of Postmodernity*, London/New York: Routledge.

Beck, U., A. Giddens and S. Lash (1994), *Reflexive Modernization. Politics, Tradition and Aesthetics in the Modern Social Order*, Cambridge: Polity Press.

Bekke, A.J.G.M., J.L.M. Hakvoort and J.M. de Heer (eds) (1994), *Departementen in beweging*, 's-Gravenhage: VUGA.

Bekke, H. and P. Kuypers (1990), *Afzien van macht.* Adviseren aan een andere overheid, 's-Gravenhage: SDU.

Bekkers, V.J.J.M. (1993), *Nieuwe vormen van sturing en informatisering*, Delft: Eburon.

Bekkers, V.J.J.M. (ed.) (1994), *Wegwijs op de digitale snelweg.* Enkele politiek-bestuurlijke aspecten van de informatie-maatschappij, Amsterdam: Cramwinckel.

Bekkers, V.J.J.M. and W.B.H.J. van de Donk (1989), Van sturing naar metasturing: een verkenning van de rol van ideologie en informatietechnologie bij de ontwikkeling van nieuwe sturingsconcepties, in A.B. Ringeling, I.Th.M. Snellen (eds), *Overheid: op de (terug)tocht of op weg naar een nieuw profiel?* Alphen aan den Rijn: VUGA, pp. 149–162.

Bekkers, V.J.J.M. and P.H.A. Frissen (1992), 'Informatization and Administrative Modernization in the Netherlands', in P.H.A. Frissen, V.J.J.M. Bekkers, B.K. Brussaard, I.Th.M. Snellen and M. Wolters (eds), *European Public Administration and Informatization. A Comparative Research Project into Policies, Systems, Infrastructures and Projects,* Amsterdam/Oxford/Washington/Tokyo: IOS Press.

Bekkers, V.J.J.M., G.J. Straten, P.H.A. Frissen, P.A. Tas, H.P.M. van Duivenboden, J. Huigen and S.B. Luijtjens (1995*), De Gemeentelijke Basisadministratie.* Een tussentijdse beoordeling van een interorganisationeel informatiseringsproject, Tilburg/'s-Gravenhage/Eindhoven: Katholieke Universiteit Brabant.

Bell, D. (1979), 'The Social Framework of the Information Society', in M.L. Dertouzos, J. Moses (eds.), *The Computer Age:* A Twenty-Year View. Cambridge/London: M.I.T. Press, pp. 163–211.

Bell, D. (1973), *The Coming of Post-Industrial Society.* A Venture in Social Forecasting, New York: Basic Books.

Bell, D. (1976), *The Cultural Contradictions of Capitalism*, New York: Basic Books.

Bell, D. (1960), *The End of Ideology,* Glencoe, Illinois: The Free Press.

Bellamy, C. and J.A. Taylor (eds) (1994), 'Towards the Information Policy? Public Administration in the Information Age', *Public Administration,* 72 (1994), **1**.

Benedikt, M. (ed.) (1991), *Cyberspace:* First Steps, Cambridge/London: MIT.

Beniger, J.R. (1986), *The Control Revolution*, Cambridge: Harvard University Press.

Berenschot (1995), *De electronische snelweg.* Een routeplanner voor managers. Alphen aan den Rijn: Samsom BedrijfsInformatie.

Bergquist, W. (1993), *The Postmodern Organization.* Mastering the Art of Irreversible Change, San Francisco: Jossey-Bass.

Beunders, H.J.B. (1994), *De strijd om het beeld.* Over de behoefte aan censuur, 's-Gravenhage: Delwel.

Beus, J. de (1993), *Economische gelijkheid en het goede leven,* Amsterdam: Contact.

Bijker, W.E. and J. Law (eds) (1992*)*, *Shaping Technology, Building Society*: Studies in Sociotechnical Change, Cambridge/London: MIT Press.

Bolkestein, F. (1992), *Woorden hebben hun betekenis*, Amsterdam: Prometheus.

Boutellier, J.C.J. (1994), 'De zorgzame staat. Over het morele motief in overheidsbeleid', *Justitiële verkenningen*, 20 (1994) **6**, 85–101.

Bovens, M.A.P. (1988), 'De rechtvaardiging van beginselen van rechtvaardigheid', *Acta Politica*, XXII (1988) **3**, 333–357.

Bovens, M., W. Derksen, W. Witteveen, P. Kalma and F. Becker (1994), 'Den Haag heeft steeds minder te vertellen', *NRC Handelsblad*, 19 September 1994.

Bovens, M., W. Derksen, W. Witteveen, F. Becker and P. Kalma (1995), *De verplaatsing van de politiek*. Een agenda voor democratische vernieuwing, Amsterdam: Wiarda Beckman Stichting.

Bovens, M., M. Trappenburg and W. Witteveen (1994), 'Voor God en Vaderland? Communitarisme en republikanisme in Nederland', *Socialisme & Democratie*, 51 (1994) **7/8**, 322–331.

Boxum, J.L., J. de Ridder and M. Scheltema (1989), *Zelfstandige bestuursorganen in soorten*, Deventer: Kluwer.

Brinckmann, H. and S. Kuhlmann (1990*)*, *Computerbürokratie*. Ergebnisse von 30 Jahren öffentlicher Verwaltung mit Informationstechnik, Opladen: Westdeutscher Verlag.

Bruijn, J.A. de and E.F. ten Heuvelhof (1991), *Sturingsinstrumenten voor de overheid.* Over Complexe netwerken en een tweede generatie sturingsinstrumenten, Leiden/Antwerpen: Stenfert Kroese.

Brussaard, B.K. (1989), 'Coördinatie van informatievoorziening. Nodig, maar ook mogelijk?', *Bestuur*, 8 (1989) **5**, 134–138.

Brussaard, B.K. (1992), 'Large Scale Information Systems: A Comparative Analysis', in P.H.A. Frissen, V.J.J.M. Bekkers, B.K. Brussaard, I.Th.M. Snellen and M. Wolters (eds), *European Public Administration and Informatization.* A Comparative Research Project into Policies, Systems, Infrastructures and Projects, Amsterdam/Oxford/Washington/Tokyo: IOS Press, pp. 171–183.

Burgers, J.P.L. (1989), 'Ruimte voor individualisering. Notities over de ruimtelijke context van sociale betrekkingen', *Beleid & Maatschappij*, (1989) **6**, 313–321.

Burrell, G. (1988), 'Modernism, Post-Modernism and Organizational Analysis 2: The contribution of Michel Foucault', *Organization Studies*, 9 (1988) **2**, 221–235.

Burrell, G. (1993), 'Eco and the Bunnymen', in J. Hassard and M. Parker (eds), *Postmodernism and Organizations*, London/Newbury Park/New Delhi: Sage Publications, pp. 83–100.

Burrows, R. (1995), 'Cyberpunk and Social and Political Theory', *The Governance of Cyberspace Conference*, Teesside.

Callinicos, A. (1989), *Against Postmodernism:* A Marxist Critique, Cambridge/Oxford: Polity Press.

Castells, M. (1996–1998), *The information age*: economy, society and culture (3 Volumes), Cambridge, MA: Blackwell.

Clegg, S.R. (1990), *Modern Organizations.* Organization studies in the Postmodern World, London/Newbury Park/New Dehli: Sage.

Commissie-De Koning (1993), *Het bestel bijgesteld*, Tweede Kamer, vergaderjaar 1992–1993, 21427, nrs 36–37.

Commissie-Scheltema (1993), *Steekhoudend ministerschap*, Betekenis en toepassing van de ministeriële verantwoordelijkheid, Tweede Kamer, vergaderjaar 1992–1993, 21427, nrs 40–41.

Commissie-Sint (1994), *Verantwoord verzelfstandigen*, 's-Gravenhage: Ministerie van Binnenlandse Zaken.

Commissie-Van Thijn (1993), *De burgemeester ontketend*, Tweede Kamer, vergaderjaar 1992–1993, 21427, nrs 34–35.

Commissie-Wiegel (1993*), Naar kerndepartementen.* Kiezen voor een hoogwaardige en flexibele rijksdienst, Tweede Kamer, vergaderjaar 1992–1993, 21427, nr 52.

Coolen, M. (1992*), De machine voorbij*. Over het zelfbegrip van de mens in het tijdperk van de informatietechniek, Meppel: Boom.

Cooper, R. (1989), 'Modernism, Postmodernism and Organizational Analysis 3: The Contribution of Jacques Derida', *Organization Studies*, 10 (1989) **4**, 477–502.

Cooper, R., G. Burrell (1988), 'Modernism, Postmodernism and Organizational Analysis: an Introduction', *Organization Studies*, 9 (1988) **1**, 91–112.

Crary, J. and S. Kwinter (eds) (1992), *Incorporations*, New York: Zone.

Crichton, M. (1994), *Disclosure*: A Novel, London: Century.

Crook, S., J. Pakulski and M. Waters (1992), *Postmodernization.* Change in Advanced Society, London/Newbury park/New Delhi: Sage.

Crozier, M. (1963), *La phénomène bureaucratique*, Paris: Seuil.

Daalder, H. (1990), Consociationalism, Centre and Periphery in the Netherlands, in H. Daalder, *Politiek en historie*. Opstellen over politiek en vergelijkende politieke wetenschap, Amsterdam: Bakker, pp. 21–63.

Dam, M.P.A. van (1992), 'De staat van de Staat', *De Volkskrant*, 31 december 1992.

Danziger, J.N., W.H. Dutton, R. Kling and K.L. Kraemer (1982), *Computers and Politics*. High Technology in American Local Governments, New York: Columbia University Press.

Davenport, T.H. (1993), *Process Innovation*: Reengineering Work through Information Technology, Boston, Massachusetts: Harvard Business School Press.

Deal, T.E. and A.A. Kennedy (1982), *Corporate Cultures*. The Rites and Rituals of Corporate Life, Reading: Addison-Wesley Pub. Co.

Deleuze, G. (1992), *Het denken in plooien geschikt*, Kampen: Kok Agora.

Deleuze, G. and F. Guattari (1977), *Rhizom*, Berlin: Merve.

Depla, P. (1995), *Technologie en de vernieuwing van de lokale democratie. Vervolmaking of vermaatschappelijking*, 's-Gravenhage: VUGA.

Depla, P. and J. Monasch (1994), *In de buurt van politiek*. Handboek voor vernieuwing van de lokale democratie, Amsterdam: Wiardi Beckman Stichting/Centrum voor Lokaal Bestuur.

Depla, P. and P.W. Tops (1992), 'Nieuwe technologie impuls voor vernieuwing bestuur', *Staatscourant*, 23 July 1992: 2–7

Derrida, J. (1974), *Éperons: les styles de Nietzsche*, Paris: Flammarion.

Derrida, J. (1981), *Glas*. Que reste-t-il du savoir absolu? 1 and 2, Paris: Denoël.

Dickson, D. (1990), *Het verval van de geest*. De technologische cultuur in het post-moderne Europa, Amsterdam: De Balie.

Donk, W.B.H.J. van de, P.H.A. Frissen and I.Th.M. Snellen (1990), 'Spanningen tussen wetgeving en systeemontwikkeling: De Wet Studiefinanciering', *Beleidswetenschap*, 4 (1990) 1, 3–20.

Donk, W.B.H.J. van de and P.H.A. Frissen (1994), 'Informatisering, sturing en wetgeving', in Ph. Eijlander, P.H.A. Frissen, P.C. Gilhuis, J.H. van Kreveld and B.W.N. de Waard (eds), *Wetgeven en de maat van de tijd*, Zwolle: W.E.J. Tjeenk Willink, pp. 44–64.

Donk, W.B.H.J. van de, I.Th.M. Snellen and P.W. Tops (eds) (1995), *Orwell in Athens*. A Perspective on Informatization and Democracy, Amsterdam/Oxford/Tokyo/Washington: IOS Press.

Donk, W.B.H.J. van de and P.W. Tops (1992), 'Informatisering en democratie: Orwell of Athene?', P.H.A. Frissen, A. Koers and I.Th.M. Snellen (eds), *Orwell of Athene? Democratie en informatiesamenleving*, 's-Gravenhage: Sdu Juridische & Fiscale Uitgeverij, pp. 31–74.

Duivenboden, H. van (1994), 'Achter de schermen van de overheid: beleid en recht inzake koppeling van persoonsregistraties', in A. Zuurmond, J. Huigen, P.H.A. Frissen, I.Th.M. Snellen and P.W. Tops (eds) *Informatisering in het openbaar bestuur*. Technologie, politiek en sturing, bestuurskundig beschouwd, 's-Gravenhage: VUGA, pp. 397–414.

Dunk, H.W. von der (1993), 'Nederland wordt een fluwelen regelstaat', *NRC Handelsblad*, 7 september 1993.

Edwards, A. (1993), 'De 'missing link' in het onderzoeksveld', *Zeno*, 1 (1993) 6, 27–29.

Eijlander, Ph., P.C. Gilhuis and J.A.F. Peters (eds) (1993), *Overheid en zelfregulering*. Alibi voor vrijblijvendheid of prikkel tot aktie? Zwolle: W.E.J. Tjeenk Willink.

Eijlander, Ph., P.H.A. Frissen, P.C. Gilhuis, J.H. van Kreveld and B.W.N. de Waard (eds) (1994), *Wetgeven en de maat van de tijd*, Zwolle: W.E.J. Tjeenk Willink.

Enzensberger, H.M. (1990), *Lof van de inconsequentie*, Amsterdam: De Bezige Bij.

Ettighoffer, D. (1993), 'L'entreprise virtuelle et les nouveaux modes d'emploi au XXIe siècle', *Nouvelles Technologies de l'Information et de l'Economie et de la Société*, Rencontres des 9 et 10 décembre 1993, Futuroscope de Poitiers.

Featherstone, M. (1988), 'In Pursuit of the Postmodern: An Introduction'. *Theory, Culture & Society*, 5 (1988) **2–3**, 195–216.

Feldman, M.S. and J.G. March (1981), 'Information in Organizations as Signal and Symbol', *Administrative Science Quarterly*, 26 (1981), 171–186.

Feldman, M.S. and M. Sarbough-Thompson (1993), 'Electronic Communication and Decision Making', *EGPA-Conference 1993*, Strasbourg.

Fischer, F. (1990), *Technocracy and the Politics of Expertise*, Newbury Park/London/New Delhi: Sage.

Foqué, R. (1992), *De ruimte van het recht*, Arnhem: Gouda Quint.

Fortuyn, W.S.P. (1992), *Aan het volk van Nederland*. De contract-maatschappij, een politiek-economische zedenschets, Amsterdam/Antwerpen: Contact.

Foster, H. (ed.) (1985), *Postmodern Culture*, London: Pluto Press.

Foster, H. (1985), 'Postmodernism: A Preface', in H. Foster (ed.), *Postmodern Culture*, London: Pluto Press, pp. IX–XVI.

Fox, C.J. and H.T. Miller (1997), *Postmodern public administration*: toward discourse, Thousand Oaks: Sage.

Frissen, P.H.A. (1989), *Bureaucratische cultuur en informatisering*. Een studie naar de betekenis van informatisering voor de cultuur van een overheidsorganisatie, 's-Gravenhage: Sdu.

Frissen, P.H.A. (1990), 'Besturingsconcepties, recht en wetgeving', in H.A.M. Backx and E.M.H. Hirsch Ballin (eds), *Recht doen door wetgeving*. Opstellen over wetgevingsvraagstukken, Zwolle: W.E.J. Tjeenk Willink, pp. 13–29.

Frissen, P.H.A. (1991), *De versplinterde staat*. Over informatisering, bureaucratie en technocratie voorbij de politiek, Alphen aan den Rijn: Samsom H.D. Tjeenk Willink.

Frissen, P.H.A.(1992[a]), 'Informatisering en bestuurlijke vernieuwing', in T. Huppes (ed.), *Informatisering en de kwaliteit van bestuur en samenleving*, Deventer: Kluwer Bedrijfswetenschappen, pp. 43–59.

Frissen, P.H.A. (1992[b]), 'Digitale dubbelzinnigheden', in P.H.A. Frissen, A. Koers and I.Th.M. Snellen (eds), *Orwell of Athene?* Democratie en informatiesamenleving, 's-Gravenhage: Sdu Juridische & Fiscale Uitgeverij, pp.179–197.

Frissen, P.H.A. (1992[c]), 'Symposium on information in public administration', *International Review of Administrative Sciences*, 58 (1992), **3**.

Frissen, P.H.A. (1993[a]), 'De versplinterde staat', *Sociale Wetenschappen*, 36 (1993) **2**, 1–22.

Frissen, P.H.A. (1993[b]), 'Zelfregulering en besturingsconcepties. Enkele bestuurskundige opmerkingen', in Ph. Eijlander, P.C. Gilhuis and J.A.F. Peters (eds), *Overheid en zelfregulering*. Alibi voor vrijblijvendheid of prikkel tot aktie? Zwolle: W.E.J. Tjeenk Willink, pp. 169–176.

Frissen, P.H.A. (1993[c]), 'Informatiemaatschappij en parlementaire democratie', in Ministerie van Economische Zaken, *Vooruitkijken naar vooruitgaan*. Technologie in de toekomst. 's-Gravenhage: Ministerie van Economische Zaken, pp. 119–129.

Frissen, P.H.A. (1993[d]), 'De smalle marges van de politieke vernieuwing', *De Helling*, 6 (1993) **3**, 32–35.

Frissen, P.H.A. (1993[e]), 'Over voorzichtigheid en verbeeldingskracht', *Socialisme & Democratie*, 50 (1993) **10**, 425–426.

Frissen, P.H.A. (1994[a]), 'De calculerende burger en de na-calculerende staat. De virtuele werkelijkheid van informatisering in het openbaar bestuur', *Bestuurskunde*, 3 (1994) **5**, 209–217.

Frissen, P.H.A. (1994[b]), 'De virtuele werkelijkheid van informatisering in het openbaar bestuur', in A. Zuurmond, J. Huigen, P.H.A. Frissen, I.Th.M. Snellen and P.W. Tops (eds), *Informatisering in het openbaar bestuur*. Technologie, politiek en sturing, bestuurskundig beschouwd, 's-Gravenhage: VUGA.

Frissen, P.H.A. (1994[c]), 'Informatisering en regionalisering', in W.G.M. Salet and H. Stevens (eds), *Gedifferentieerd Regionaal Bestuur*, Delft: Faculteit der Technische Bestuurskunde, TU Delft, pp. 39–48.

Frissen, P.H.A. (1994[d]), 'Fragmentatie in een technologische cultuur. Individualisering en solidariteit in bestuurskundig perspectief', in A. van den Broek and B. Seuren (eds), *Individualisering & Solidariteit*, Tilburg: Tilburg University Press, pp. 19–30.

Frissen, P.H.A., (1994^c), 'The virtual reality of informatization in public administration', *Informatization and the Public Sector*, 3 (1994) **3/4**, 265–291.

Frissen, P.H.A., P. Albers, V.J.J.M. Bekkers, J. Huigen, K. Schmitt, M. Thaens and B. de Zwaan (1992^a), *Verzelfstandiging in het openbaar bestuur*. Een bestuurskundige verkenning van verzelfstandiging, verbindingen en informatisering, 's-Gravenhage: VUGA.

Frissen, P.H.A., V.J.J.M. Bekkers, B.K. Brussaard, I.Th.M. Snellen and M. Wolters (eds) (1992^b), *European Public Administration and Informatization*. A Comparative Research Project into Policies, Systems, Infrastructures and Projects, Amsterdam/Oxford/Washinton/Tokyo: IOS Press.

Frissen, P.H.A., E.M.H. Hirsch Ballin, R.J. Hoekstra, R.K. Visser, R.J. Kuiper and R. de Boer (1993), *Het managen van verandering door verzelfstandiging*, 's-Gravenhage: Management Centrum.

Frissen, P.H.A., A. Koers and I.Th.M. Snellen (1992), *Orwell of Athene?* Democratie en informatiesamenleving, 's-Gravenhage: Sdu Juridische & Fiscale Uitgeverij.

Frissen, P.H.A. and I.Th.M. Snellen (eds.) (1990), *Informatization Strategies in Public Administration*. Informatization developments and the Public Sector, I, Amsterdam/New York/Oxford/Tokyo: IOS Press.

Fruytier, B. (1994), *Organisatieverandering en het probleem van de Baron van Münchhausen*. Een systeemtheoretische analyse van de overgang van het Tayloristisch Produktie Concept naar het Nieuwe Produktie Concept, Delft: Eburon.

Fukuyama, F. (1989), 'The End of History?', *The National Interest*, 1989, summer, 3–18.

Geertsema, H.G. (1988), *Hoe kan de wetenschap menselijk zijn?* Amsterdam: VU Uitgeverij.

Gellner, E. (1992), *Postmodernism, Reason and Religion*, London: Routledge.

Geurts, J.L.A. (1993), *Omkijken naar de toekomst*. Lange termijn verkenningen in beleidsexcercities, Alphen aan den Rijn: Samsom H.D. Tjeenk Willink.

Geus, M. de (1989), *Organisatietheorie en de politieke filosofie*, Delft: Eburon.

Gibbins, J.R. (ed.) (1989), *Contemporary Political Culture*. Politics in a Postmodern Age, London/Newbury Park/ New Delhi: Sage.

Gibbins, J.R. (1989), 'Contemporary Political Culture: an Introduction', in J.R. Gibbins (ed.), *Contemporary Political Culture*. Politics in a Postmodern Age, London/Newbury Park/ New Delhi: Sage, pp. 1–27.

Giddens, A. (1990), *The Consequences of Modernity*, Cambridge: Polity Press.

Giddens, A. (1991), *Modernity and Self-Identity*. Self and Society in the Late Modern Age, Cambridge: Polity Press.

Godfroij, A.J.A. and N.J.M. Nelissen (eds) (1993), *Verschuivingen in de besturing van de samenleving*, Bussum: Couthino.

Gooren, W.A.J. and B.C. de Zwaan (1993), *De reorganisatie van de organisatie*. Verslag van een onderzoek naar de invoering van het nieuwe politiebestel in de korpsen, Tilburg: IVA.

Guéhenno, J.-M. (1994), *Het einde van de democratie*, Utrecht: Lannoo.

Gunsteren, H.R. van (1984), 'Wie in zijn graf ligt, maakt geen fouten meer: een interventieleer te ontwikkelen door ambtenaren', *Beleid & Maatschappij*, XI (1984) **6**, 159–163.

Gunsteren, H.R. van (1993), 'Eenvoud in veelvoud', *Beleid & Maatschappij*, XX (1993) **1**, 3–7.

Gunsteren, H.R. van (1994), 'Culturen van calculatie', *Bestuurskunde*, 3 (1994) **5**, 218–226.

Gunsteren, H.R. van and R. Andeweg (1994), *Het grote ongenoegen*. Over de kloof tussen burgers en politiek, Haarlem: Aramith.

Gunsteren, H.R. van and E. van Ruyven (1993), 'De Ongekende Samenleving (DOS), een verkenning', *Beleid & Maatschappij*, XX (1993) **3**, 114–125.

Haan, I. de (1993), *Zelfbestuur en staatsbeheer*. Het politieke debat over burgerschap en rechtsstaat in de twintigste eeuw, Amsterdam: Amsterdam University Press.

Habermas, J. (1985[a]), *Die Neue Unübersichtlichkeit*. Kleine Politische Schriften V, Frankfurt am Main: Suhrkamp.

Habermas, J. (1985[b]), 'Modernity – An Incomplete Project', in H. Foster (ed.), *Postmodern Culture*, London: Pluto Press, pp. 3–15.

Habermas, J. (1987), *The Philosophical Discourse of Modernity*. Twelve Lectures, Cambridge: Polity in association with Basil Blackwell.

Harvey, D. (1989), *The Condition of Postmodernity*. An Inquiry into the Origins of Cultural Change, Cambridge/Oxford: Blackwell.

Hassan, I. (1985), 'The Culture of Postmodernism', *Theory, Culture and Society*, 2 (1985) **3**, 119–132.

Hassard, J. and M. Parker (eds) (1993), *Postmodernism and Organizations*, London/Newbury Park/New Delhi: Sage Publications.

Heidegger, M. (1954), 'Die Frage nach der Technik', in M. Heidegger, *Vorträge und Aufsätze*, Teil I, II, III, Pfullingen: Neske, pp. 5–36.

Heidegger, M. (1967), *Vorträge und Aufsätze*, Teil I, II, III, Pfullingen: Neske.

Heller, A. and F. Fehér (1988), *The Postmodern Political Condition*, Cambridge: Polity Press.

Hirsch Ballin, E.M.H. (1988), 'De christen-democratische politieke overtuiging omtrent de reikwijdte van democratisch-rechtstatelijke optiek', in A.M.J.

Kreukels and J.B.D. Simonis (eds), *Publiek domein*: de veranderende balans tussen staat en samenleving, Meppel: Boom, pp. 111–137.

Hirsch Ballin, E.M.H. (1991), 'De gekoppelde staat', in L.A. Geelhoed (e.a.), *Wetgeving in beweging,* Zwolle: W.E.J. Tjeenk Willink, pp. 61–74.

Hirsch Ballin, E.M.H. (1993), *De staat van Nederland*, Rede, Tilburg: Tilburg University Press.

Hirsch Ballin, E.M.H. (1994^a), 'Communitarisme voor de Nederlandse samenleving', *Socialisme & Democratie*, 51 (1994) **7/8**, 339–344.

Hirsch Ballin, E.M.H. (1994^b), *In ernst*. Oriëntaties voor beleid, 's-Gravenhage: SDU Juridische & Fiscale Uitgeverij.

Hirsch Ballin, E.M.H. (1995), 'Echte politici laten zich door partijgrenzen niet weerhouden', *NRC Handelsblad*, 2 september 1995.

Hoed, P. den, W.G.M. Salet and M. van der Sluijs (1983), *Planning als onderneming*, 's-Gravenhage: Staatsuitgeverij.

Hoff, J. and K. Stormgaard (1990), 'A Reinforcement Strategy for Informatization in Public Administration in Denmark?', in P.H.A. Frissen and I.Th.M. snellen (eds.), *Informatization Strategies in Public Administration*. Informatization developments and the Public Sector, I, Amsterdam/New York/Oxford/Tokyo: IOS Press, pp. 107–132.

Holthoorn, F. van (1988), 'De geschiedenis van het publiek domein in Nederland sinds 1815', in A.M.J. Kreukels and J.B.D. Simonis (eds), *Publiek domein*: de veranderende balans tussen staat en samenleving, Meppel: Boom, pp. 57–85.

Hood, Chr. and H. Margetts (1993), 'Informatization and Public Administration Trends', *ESCR/PICT Seminar on Public Sector Informatization*, London.

Hoogerwerf, A. (1995), *Politiek als evenwichtskunst*. Dilemma's rond overheid en markt, Alphen aan den Rijn: Samsom H.D. Tjeenk Willink.

Hoppe, R. and A. Edwards (1985), 'Beleidsvorming: benaderingen in soorten en maten', in R. Hoppe (ed.), *Trends in beleidsvormingstheorie en ontwerpleer*. Amsterdam: VU Uitgeverij.

Hoven, M.J. van den (1994), 'Towards ethical principles for designing politico-administrative information systems', *Informatization and the Public Sector* 3 (1994) **3/4**, 353–373.

Huigen, J. (1994), *Information Supply and the Implementation of Policy*. Playing with Ambiguity and Uncertainty in Policy Networks, Delft: Eburon.

Huigen, J., P.H.A. Frissen and P.W. Tops (1993), *Het project Betuweroute: spoorlijn of bestuurlijke co-produktie?* Leerervaringen voor besluitvorming inzake grootschalige infrastructuur, Tilburg: CRBI.

Huigen, J., M. Thaens and P.H.A. Frissen (1994), *Schatting van baten van informatiesystemen bij de Belastingdienst.* Organisatorisch concept, Tilburg: CRBI.

Hummel, R.P. (1990), 'Circle Managers and Pyramid Managers: Icons for the Post-Modern Public Administration', in H.D. Kass and B.L. Catron (eds), *Images and Identities in Public Administration,* Newbury Park/London/New Delhi: Sage, pp. 202-218.

Hupe, P.L. (1995), 'Het betwiste primaat van de politiek', in P. de Jong, A.F.A. Korsten, A.J. Modderkolk and I.M.A.M. Pröpper (eds), *Verantwoordelijkheid en verantwoording in het openbaar bestuur,* 's-Gravenhage: VUGA, pp. 61-70.

Huyse, L. (1994), *De politiek voorbij.* Een blik op de jaren negentig, Leuven: Kritak.

Idenburg, Ph.A. and H.R. van der Loo (1993), 'De staarten van de rattenkoning: opmerkingen over de vervlechting van wetenschap, politiek en bureaucratie', *Beleidswetenschap,* 7 (1993) **2**, 124-140.

Idenburg, Ph.A. and H.R. van der Loo (1994), *In alle staten.* Een beschouwing over de rol en de betekenis van overheidsbeelden, 's-Gravenhage: VUGA.

Inbar, M. (1979), *Routine Decision-Making.* The Future of Bureaucracy, Beverly Hills/London: Sage Publications.

Jackson, N. and P. Carter (1992), 'Postmodern Management. Post-Perfect or Future-Imperfect?', *International Studies of Management and Organization,* 22 (1992) **3**, 11-26.

Jameson, F. (1984[a]), 'The Politics of Theory: Ideological Positions in the Postmodernism Debate', *New German Critique,* 33 (1984), 53-65.

Jameson, F. (1984[b]), 'Postmodernism, or the Cultural Logic of Late Capitalism', *New Left Review,* 146 (1984), 53-92.

Jencks, Ch. (1986),. *What is Post-Modernism?,* London/New York: Academy.

Jong, W.M. de (1994), *The Management of Informatization.* A theoretical and empirical analysis of IT implementation strategies, Groningen: Wolters-Noordhoff.

Jorritsma-Mientjes, T. and P.H.A. Frissen (1988), 'Cultuurverandering voor managers verklaard', *Opleiding & Ontwikkeling,* 1 (1988) **12**, 8-10.

Kass, H.D. and B.L. Catron (eds) (1990), *Images and Identities in Public Administration,* Newbury Park/London/New Delhi: Sage.

Kellner, D. (1988), 'Postmodernism as Social Theory: Some Challenges and Problems', *Theory, Culture & Society,* 35 (1988) **2-3**, 239-270.

Kelly, K. (1994), *Out of Control.* The New Biology of Machines, London: Fourth Estate.

Kickert, W.J.M. (1991), *Complexiteit, zelfsturing en dynamiek*. Over management van complexe netwerken bij de overheid, Alphen aan den Rijn: Samsom H.D. Tjeenk Willink.

Kickert, W.J.M., N.P. Mol and A. Sorber (eds) (1992), *Verzelfstandiging van overheidsdiensten*, 's-Gravenhage: VUGA.

Klink, B. van, P. van Seters, and W. Witteveen (eds) (1993), *Gedeelde normen?* Gemeenschapsdenken en het recht, Zwolle: W.E.J. Tjeenk Willink.

Klop, C.J. (1993), *De cultuurpolitieke paradox*. Noodzaak èn onwenselijkheid van overheidsinvloed op normen en waarden, Kampen: Kok.

Klop, C.J. (1994), 'Het verhaal van een republikein', *Socialisme & Democratie*, 51 (1994) **7/8**, 332–334.

Kooiman, J. (ed.) (1993), *Modern Governance*. New Government-Society Interactions, London/Newbury Park/New Delhi: Sage.

Korsten, A.F.A. and E.H.A. Willems (1993), 'Staatkundige, bestuurlijke en staatsrechtelijke vernieuwing. Rolt het ei opnieuw van tafel?', *Bestuurskunde*, (1993) **5**, 226–230.

Kreukels, A.M.J. and J.B.D. Simonis (eds) (1988), *Publiek domein*: de veranderende balans tussen staat en samenleving, Meppel: Boom.

Kuiper, R.J. (1993), 'Verzelfstandiging en informatievoorziening', in P.H.A. Frissen, E.M.H. Hirsch Ballin, R.J. Hoekstra, R.K. Visser, R.J. Kuiper and R. de Boer, *Het managen van verandering door verzelfstandiging*, 's-Gravenhage: Management Centrum.

Kuitenbrouwer, J. (1990), *Lijfstijl*. De manieren van nu, Amsterdam: Prometheus.

Kunneman, H. (1988), 'De betekenis en de beperkingen van het postmodernisme als politieke filosofie', *Socialisme & Democratie*, 45 (1988) **7/8**, 201–213.

Kuypers, P. (1994), 'De orde van de wet. Over de ambtelijke moraal', *Justitiële Verkenningen*, 20(1994) **6**, 116–122.

Kuypers, P., R. Foqué and P. Frissen (1993), *De lege plek van de macht*. Over bestuurlijke vernieuwing en de veranderende rol van de politiek, Amsterdam: De Balie.

Larrain, J. (1994), 'The Postmodern Critique of Ideology', *Sociological Review*, 42 (1994) **1**, 289–314.

Lash, S. (1990), *Sociology of Postmodernism*, London/New York: Routledge.

Lash, S. and J. Urry (1987), *The End of Organized Capitalism*, Cambridge: Polity Press, in association with Basil Blackwell.

Lenk, K. (1994), 'Information systems in public administration: From research to design', *Informatization and the Public Sector*, 3 (1994) **3/4**, 305–324.

Levy, S. (1992), *Artificial Life. The Quest for a New Creation*, New York: Pantheon Books.

Lijphart, A. (1968), *Verzuiling, pacificatie en kentering in de Nederlandse politiek*, Amsterdam: De Bussy.

Loo, H.R. van der (1994), 'De nieuwe huiselijkheid', *De Volkskrant*, 31 December 1994.

Loo, H.R. van der and Ph.A. Idenburg (1994), 'Voorbij de vanzelfsprekendheid van tijd. Implicaties van een reflexieve omgang met tijd', *Bestuurskunde*, 3 (1994) **7**, 278–294.

Lyotard, J.-F. (1987[a]), *Het postmoderne weten*, Kampen: Kok Agora.

Lyotard, J.-F. (1987[b]), *Het postmoderne uitgelegd aan onze kinderen*, Kampen: Kok Agora.

Mak, G. (1994), 'Een computerdorp voor seropositieven', *NRC Handelsblad*, 24 May 1994.

Malone, Th.E., J. Yates and R.I. Benjamin (1987), 'Electronic Markets and Electronic Hierarchies', *Communications of the ACM*, 30 (1987) **3**, 484–497.

March, J.G., J.P. Olsen (1983), 'Organizing Political Life: What Administrative Reorganization Tells Us About Government', *The American Politial Science Review*, 77 (1983) **2**, 281–296.

Marx, K. (1956), Zur Kritik der Hegelschen Rechtsphilosophie, reprinted in K. Marx and F. Engels (1977), *Werke*, Berlin: Dietz, pp. 201–333.

Mascarpone (1992), *Gebroken wit*. Politiek van de kleine verhalen, Amsterdam: Ravijn.

Meer, F.B. van der and H. Boer (1994), 'Organisatie: theoretische perspectieven', in A. Zuurmond, J. Huigen, P.H.A. Frissen, I.Th.M. Snellen and P.W. Tops (eds), *Informatisering in het openbaar bestuur*. Technologie, politiek en sturing, bestuurskundig beschouwd, 's-Gravenhage: VUGA, pp. 191–205.

Meer, F.B. van der and T. Roodink (1991), 'The dynamics of automation: a structural constructionist approach', *Informatization and the Public Sector*, 1 (1991) **2**, 121–141.

Meyer, O.M.T. (1994), 'Informatisering en politieke agendavorming: de agendavormende waarde van informatiesystemen', in A. Zuurmond, J. Huigen, P.H.A. Frissen, I.Th.M. Snellen and P.W. Tops (eds), *Informatisering in het openbaar bestuur*. Technologie, politiek en sturing, bestuurskundig beschouwd, 's-Gravenhage: VUGA, pp. 113–133.

Mieras, M. (1994[a]), 'Het pragmatisme van Internet', *Intermediair*, 30 (1994) **10**, 26–27.

Mieras, M. (1994[b]), 'Nog niet gebouwd, al wel te bezichtigen', *Intermediair*, 30 (1994) **19**, 25–27.

Mieras, M. (1994ᶜ), 'Digitaal darwinisme', *Intermediair*, 30 (1994) **34**, 35–37.

Mieras, M. (1995), 'Het net sluit zich. Koppeling van bestanden gaat steeds verder', *Intermediair*, 31 (1995) **10**, 49–51.

Ministerie van Binnenlandse Zaken (1993), *De organisatie en werkwijze van de rijksdienst.* Rapportage van de secretarissen-generaal, 's-Gravenhage: Ministerie van Binnenlandse Zaken.

Ministerie van Binnenlandse Zaken (1995), *Terug naar de toekomst.* Over het gebruik van informatie- en communicatie-technologie in de openbare sector. Beleidsnota Informatiebeleid Openbare Sector (BIOS), nr.3, 's-Gravenhage: Ministerie van Binnenlandse Zaken.

Ministerie van Economische Zaken (1994), *Nationaal Actieprogramma Electronische Snelwegen.* Van metafoor naar actie, 's-Gravenhage: Ministerie van Economische Zaken.

Mommaas, H. (1993), *Moderniteit, vrije tijd en de stad.* Sporen van maatschappelijk transformatie en continuïteit, Utrecht: Van Arkel.

Moor, M. de (1993), *De virtuoos.* Roman, Amsterdam: Contact.

Morgan, G. (1986), *Images of Organization*, Beverly Hills: Sage.

Morgan, G. (1993), *Imaginisatie.* De kunst van creatief management, Schiedam: Scriptum.

Mowshowitz, A. (1992), Virtual Feudalism: A Vision of Political Organization in the Information Age', in P.H.A. Frissen, A. Koers and I.Th.M. Snellen, *Orwell of Athene?* Democratie en informatiesamenleving, 's-Gravenhage: Sdu Juridische & Fiscale Uitgeverij, pp. 285–300.

Mulgan, G. (1994), *Politics in an Antipolitical Age*, Cambridge: Polity Press.

Naschold, F. (1993), *Modernisierung des Staates.* Zur Ordnung- und Innovationspolitik des öffentlichen Sektors, Berlin: Sigma.

Negroponte, N. (1995), *Digitaal leven*, Amsterdam: Prometheus.

Nietzsche, F. (1973), *Werke in zwei bänden.* Herausgegeben von Ivo Frenzel, Darmstadt: Wissenschaftliche Buchgesellschaft.

Nietzsche, F. (1979), *Werke IV.* Aus dem Nachlass der Achtzigerjahre. Briefe (1861–1889), Herausgegeben von Karl Schlechta, Frankfurt am main/Berlin/Wien: Ullstein.

Noordegraaf, M. (1995), 'Het bestuursraadmodel. Individualisering van structuren en postmoderne socialisatie', *Openbaar Bestuur*, 5 (1995) **6/7**, 16–19.

Nooteboom, B. (1992), 'A Postmodern Philosophy of Markets', *International Studies of Management and Organization*, 22 (1992) **2**, 53–76.

Nooteboom, B. (1993), 'Lang leve de creatieve vernietiging van ideeën', *De Volkskrant*, 23 January 1993.

Nora, S., A. Minc (1978), *Infomatisation de la société*, Paris: La Documentation Française.

Oerlemans, J.W. (1992), 'Om de macht van de vrijheid', *NRC Handelsblad*, 3 October 1992.

Offe, C. (1985), *Disorganized Capitalism*, Cambridge: Polity Press.

Palumbo, D. and S. Maynard-Moody (1991), *Contemporary Public Administration*, New York/London: Longman.

Parker, M. (1992), 'Post-Modern Organizations or Postmodern Organization Theory?', *Organization Studies*, 13 (1992) **1**, 1–17.

Perez, C. (1983), 'Structural Change and Assimilation of New Technologies in the Economic and Social Systems', *Futures*, October 1983, 357–375.

Perlee, J.A. (1993), 'Netwerken: waar de gemeente op moet letten', in J.A. van der Drift (ed.), *Gemeentelijk telecommunicatiebeleid*, I. Communicatie via computernetwerken, 's-Gravenhage: VNG, pp. 61–70.

Peters, T. (1992), *Liberation Management*. Necessary Disorganization for the Nanosecond Nineties, London: Macmillan.

Poster, M. (1990), *The Mode of Information*. Poststructuralism and Social Context, Cambridge: Polity Press.

Pot, J.H.J. van der (1985), *Die Bewertung der technische Fortschritts*. Eine systematische Übersicht der Theorie. Band I + II, Assen/Maastricht: Van Gorcum.

Reijen, W. van (1988), 'Moderne versus postmoderne politieke filosofie. Een vergelijking van Habermas en Lyotard', *Acta Politica*, XXIII (1988) **2**, 199–223.

Reiner, R. (1992), 'Policing a Postmodern Society', *The Modern Law Review*, 55 (1992) **6**, 761–781.

Rheingold, H. (1991), *Virtual Reality*, New York / London / Toronto / Sydney / Tokyo / Singapore: Simon & Schuster.

Rheingold, H. (1994), *The Virtual Community*. Finding Connection in a Computerized World, London: Minerva.

Ringeling, A. (1993), *Het imago van de overheid*. De beoordeling van prestaties van de publieke sector, 's-Gravenhage: VUGA.

Rorty, R. (1989), *Contingency, irony, and solidarity*, Cambridge: Cambridge University Press.

Rorty, R. (1991[a]), *Objectivity, Relativism and Truth*. Philosophical Papers I, Cambridge: Cambridge University Press.

Rorty, R. (1991[b]), *Essays on Heidegger and Others*. Philosophical Papers II, Cambridge: Cambridge University Press.

Rosenthal, U. (1988), *Bureaupolitiek en bureaupolitisme*. Om het behoud van een competitief overheidsbestel, Alphen aan den Rijn: Samsom H.D. Tjeenk Willink.

Rosenthal, U., H.G. Geveke and P. 't Hart (1994), 'Beslissen in een competitief overheidsbestel: bureaupolitiek en bureaupolitisme nader beschouwd', *Acta Politica*, XXIX (1994) **3**, 309–334.

Roszak, Th. (1986), *De informatiecultus*. Computerfolklore en de kunst van het denken, Amsterdam: Meulenhoff Informatief.

Salet, W.G.M. (1994), *Om recht en staat*. Een sociologische verkenning van sociale, politieke en rechtsbetrekkingen, 's-Gravenhage: Sdu.

Schalken, K. and J. Flint (1995), *Handboek Digitale Steden*, Amsterdam: Stichting de Digitale Stad.

Scheepers, A.W.A. (1991), *Informatisering en de bureaucratische competentie van de burger*, Dissertatie KUB Tilburg.

Scheltema, M. (1977), 'Raden en commissies als Zelfstandige Bestuursorganen', in W.R.R., *Adviseren aan de overheid*, 's-Gravenhage: Staatsuitgeverij.

Schmid, B. (1993), 'Elektronische Markte', *Wirtschaftinformatik*, 35 (1993) **5**, 465–480.

Schokker, T. (1994), 'Wetgeving en systeemontwikkeling', in A. Zuurmond, J. Huigen, P.H.A. Frissen, I.Th.M. Snellen and P.W. Tops (eds), *Informatisering in het openbaar bestuur*. Technologie, politiek en sturing, bestuurskundig beschouwd, 's-Gravenhage: VUGA, 147–157.

Schroeder, R. (1993), 'Virtual Reality in the Real Word: History, Applications, Projections', *Futures: a Journal of Forecasting, Planning and Policy*, 25 (1993) **9**, 963–973.

Seters, P. van (1993), 'Gemeenschapsdenken en het recht', in B. van Klink, P. van Seters, and W. Witteveen (eds), *Gedeelde normen?* Gemeenschapsdenken en het recht, Zwolle: W.E.J. Tjeenk Willink, pp.1–12.

Sherman, B. and Ph. Judkins (1992), *Glimpses of Heaven, Visions of Hell*. Virtual Reality and its Implications, London: Hodder & Stoughton.

Shusterman, R. (1988), 'Postmodernist Aestheticism: A New Moral Philosophy?', *Theory, Culture & Society*, 5 (1988) **2–3**, 337–356.

Slaats, H. and H. Knip (1992), 'Policy Making and the Organization of Learning', *International Studies of Management & Organization*, 22 (1992) **2**, 77–95.

Snellen, I.Th.M. (1987), *Boeiend en geboeid*. Ambivalenties en ambities in de bestuurskunde, Alphen aan den Rijn: Samsom H.D. Tjeenk Willink.

Snellen, I.Th.M. (1993), 'Automation of Policy Implementation', *EGPA-conference*, Strasbourg.

Snellen, I.Th.M. (1994), 'ICT: A Revolutionising Force in Public Administration?', *Informatization and the Public Sector*, 3 (1994) **3/4**, 283–304.

Snellen, I.Th.M. and W.B.H.J. van de Donk (1989), 'Some dialectical developments of informatization in public administration', *Conference on New Technologies in Public Administraion: Socio-Economic Aspects from an Interdisciplinary Viewpoint,* Zagreb.

Snellen, I.Th.M. and W.B.H.J. van de Donk (eds.) (1998), *Public Administration in an Information Age. A Handbook,* Amsterdam / Berlin / Oxford / Tokyo / Washington, DC: IOS Press / Ohmsha.

Snellen, I.Th.M. and S. Wyatt (1993), 'Blurred Partitions but Thicker Walls. Involving Citizens in Computer Supported Cooperative Work for Public Administration', *Computer Supported Cooperative Work*, 1 (1993), 277–293.

Stauth, G. and B.S. Turner (1988), 'Nostalgia, Postmodernism and the Critique of Mass Culture', *Theory, Culture & Society*, 5 (1988) **2–3**, 509–526.

Stokkom, B. van (1992), *De republiek der weerbaren*, Houten: Bohn Stafleu Van Loghum.

Stone, A.R. (1992), 'Virtual Systems', in J. Crary and S. Kwinter (eds), *Incorporations*, New York: Zone, pp. 609–621.

Stuurman, S. (1983), *Verzuiling, kapitalisme en patriarchaat*. Aspecten van de ontwikkeling van de moderne staat in Nederland, Nijmegen: SUN.

Stuurman, S. (1985), *De labyrintische staat*. Over politiek, ideologie en moderniteit, Amsterdam: SUA.

Stuurman, S. (1986), *Moderniteit en politieke theorie*, Nijmegen: SUN.

Swaan, A. de (1976), 'De mens is de mens een zorg: over verstatelijking van verzorgingsarrangementen', *De Gids*, 139 (1976) **1/2**, 35–47.

Swieringa, J. and A.F.M. Wierdsma (1990), *Op weg naar een lerende organisatie*, Groningen: Wolters Noordhoff.

Teisman, G.R. (1992), *Complexe besluitvorming*. Een pluricentrisch perspektief op besluitvorming over ruimtelijke investeringen, 's-Gravenhage: VUGA.

Teubner, G. (1983), 'Substantive and Reflexive Elements in Modern Law', *Modern Law Review*, 17(1983) **2**, 239–285.

Thompson, O. (1993), 'Postmodernism: Fatal Distraction', in J. Hassard and M. Parker (eds), *Postmodernism and Organizations*, London/Newbury Park/New Delhi: Sage Publications, pp.183–203.

Tjeenk Willink, H.D. (1994), *Democratie als Beeldenstrijd*, 's-Gravenhage: VUGA.

Tjeenk Willink, H.D., W. Derksen, M. Oosting and J.A. Nekkers (1991), *De Bestuurlijke vernieuwing en de Partij van de Arbeid als bestuurderspartij*, Amsterdam: Wiardi Beckman Stichting.

Tops, P.W. (1994), *Moderne regenten*. Over lokale democratie, Amsterdam/Antwerpen: Atlas.

Tops, P.W. (1995), *Gemeenten en gezag*. Het verschuivende politieke moment in het lokale bestuur, Amsterdam/Antwerpen: Atlas.

Tops, P.W., S.A.H. Denters, P. Depla, J.W. van Deth, M.H. Leijenaar and B. Niemöller (1991), *Lokale democratie en bestuurlijke vernieuwing in Amsterdam, Den Haag, Utrecht, Eindhoven, Tilburg, Nijmegen en Zwolle*, Delft: Eburon.

Tops, P.W., P. Depla and A.F.A. Korsten (eds) (1993), Lokale bestuurlijke vernieuwing en nieuwe technologie, *Bestuurskunde*, 2 (1993) **6**.

Tromp, B. (1990), *Het einde van de politiek?* Schoonhoven: Academic Service.

Turkle, S. (1984), *The Second Self*. Computers and the Human Spirit, New York: Simon and Schuster.

Turner, B.S. (1989), 'From Postindustrial Society to Postmodern Politics: the Political Sociology of Daniel Bell', in J.R. Gibbins (ed.), *Contemporary Political Culture*. Politics in a Postmodern Age, London/Newbury Park/ New Delhi: Sage, pp. 199–217.

Twist, M.J.W. van (1995), *Verbale vernieuwing*. Aantekeningen over de kunst van bestuurskunde, 's-Gravenhage: VUGA.

Twist, M.J.W. van and R.J. in 't Veld (1994), 'Een kerndepartement is iets anders dan wat er na verzelfstandiging overblijft', *Openbaar Bestuur*, 4 (1994) **8**, 14–19.

Twist, M..J.W. van and R.J. in 't Veld (1995), 'Het kerndepartement als sterfhuis. Over de onzekere toekomst van 's rijks ministeries', *Bestuurskunde*, 4 (1995) **5**, 201–209.

Veld, R.J. in 't (1984), *De vlucht naar Isfahan?* Over bestuur, planning en de toekomst van het hoger onderwijs, 's-Gravenhage: VUGA.

Veld, R.J. in 't (1993[a]), Door de spiegels van utopie en waarheid. Een beschouwing over de toekomstige rijksoverheid vanuit het lokaal bestuur, *Jaarbericht 1993/94*, Amsterdam: Wiarda Beckman Stichting.

Veld, R.J. in 't (1993[b]), 'Zelfregulering en overheidssturing', in Ph. Eijlander, P.C. Gilhuis and J.A.F. Peters (eds), *Overheid en zelfregulering*. Alibi voor vrijblijvendheid of prikkel tot aktie? Zwolle: W.E.J. Tjeenk Willink, pp. 53–68.

Veld, R.J. in 't, P.E. Noorman and A. van der Zwan (1990), *Management in professionele organisaties*, 's-Gravenhage: SDU.

Veld, R.J. in 't, L. Schaap, C.J.A.M. Termeer, M.J.W. van Twist (eds) (1991), *Autopoiesis and Configuration Theory: New Approaches to Societal Steering*, Dordrecht/Boston/London: Kluwer Academic Publishers.

Verhallen, H.J.G., R. Fernhout and P.E. Visser (1980), *Corporatisme in Nederland: belangengroeperingen en democratie*, Alphen aan den Rijn: Samsom.

Virilio, P. (1995), Alerte dans le cyberspace! *Le Monde Diplomatique*, 42 (1995) **497**, 28.

Vries, J. de and A.F.A. Korsten (eds) (1992), 'Verzelfstandiging bij de overheid', Themanummer. *Bestuurskunde*, 1 (1992) **1**.

Vroom, C.W. (1980), *Bureaucratie, het veelzijdig instrument van de macht*. Een voorstel tot herordening van de organisatiesociologie, Alphen aan den Rijn: Samsom.

Wansink, H. (1994), 'Het ik-tijdperk heeft zijn langste tijd gehad', *NRC Handelsblad*, 17 October 1994.

Weber, M. (1968), *Economy and Society*, New York.

Weber, M. (1985), *Wirtschaft und Gesellschaft*. Grundriss der verstehenden Soziologie, Tübingen: Mohr.

White, S.K. (1991), *Political Theory and Postmodernism*, Cambridge: Cambridge University Press.

Willcocks, L. (1994), 'Managing information systems in UK public administration: issues and prospects', *Public Administration*, 72 (1994) **1**, 13–32.

Willke, H. (1992), *Ironie des Staates*. Grundlinien einer Staatstheorie polyzentrischer Gesellschaft, Frankfurt am Main: Suhrkamp.

Wit, Th.W.A. de (1995), 'De ontluistering van de politiek. Over 'Éénpartijstaat Nederland' en postmoderne democratie', *Socialisme & Democratie*, 52 (1995) **4**, 155–165.

Wöltgens, Th. (1992), *Lof van de politiek*, Amsterdam: Prometheus.

WRR (1992), *Eigentijds burgerschap*. WRR-publikatie vervaardigd onder leiding van H.R. van Gunsteren, 's-Gravenhage: Sdu.

Zeef, P. (1994), *Tussen toezien en toezicht*. Veranderingen in bestuurlijke toezichts-verhoudingen door informatisering, Den Haag: Phaedrus.

Zeef, P. and A. Zuurmond (1994), 'Structuren in viervoud', in Zuurmond, A., J. Huigen, P.H.A. Frissen, I.Th.M. Snellen and P.W. Tops (eds), *Informatisering in het openbaar bestuur*. Technologie, politiek en sturing, bestuurskundig beschouwd, 's-Gravenhage: VUGA, pp. 225–246.

Zijderveld, A.C. (1983), *De culturele factor*. Een cultuursociologische wegwijzer, 's-Gravenhage: VUGA.

Zijderveld, A.C. (1991), *Staccato Cultuur, flexibele maatschappij en verzorgende staat*. De ironie van wat ons drijft en belangrijk dunkt, Utrecht: Lemma.

Zinke, R.C. (1990), 'Administron: The Image of the Administrator as Information Processor', in H.D. Kass and B.L. Catron (eds), *Images and Identities in Public Administration*, Newbury Park/London/New Delhi: Sage, pp. 183–201.

Zolo, D. (1992), *Democracy and Complexity*. A Realist Approach, Cambridge: Polity Press.

Zouridis, S., P.H.A. Frissen and C.A.T. Schalken (1995), *Digitale participatie*. Verslag van een experimenteel digitaal debat over (de Wet) openbaarheid van bestuur, Tilburg: CRBI.

Zouridis, S. and I.Th.M. Snellen (1994), 'Informatisering en uitvoering van overheidsbeleid', in Zuurmond, A., J. Huigen, P.H.A. Frissen, I.Th.M. Snellen and P.W. Tops (eds), *Informatisering in het openbaar bestuur*. Technologie, politiek en sturing, bestuurskundig beschouwd, 's-Gravenhage: VUGA, pp. 135–146.

Zuboff, S. (1988), *In the Age of the Smart Machine*. The Future of Work and Power, Oxford: Heinemann.

Zuurmond, A. (1994), *De infocratie*. Een theoretische en empirische heroriëntatie op Weber's ideaaltype in het informatietijdperk, Den Haag: Phaedrus.

Zuurmond, A., J. Huigen, P.H.A. Frissen, I.Th.M. Snellen and P.W. Tops (eds) (1994), *Informatisering in het openbaar bestuur*. Technologie, politiek en sturing, bestuurskundig beschouwd, 's-Gravenhage: VUGA.

Index

accessibility, of information 70
administrative theory, postmodern 238–43
aesthetics 168–9, 202–7, 261–3
ambitions 80–1, 252–3
Amsterdam 69
anarchism, political postmodernism 268
anthropocentrism 60
anti-metaphysics 166–8
anti-representationalism 167
anti-utopia 258
anticipation, public administration 80–1
apolitical domain 255–8
applications, informatisation 46–50, 65–7
arbitrariness, of politics 211, 253
archipelago metaphor
   bureaucracy 156
   public administration 2, 31–2, 230–1
   society 142–3
artificial life 59–60
autonomisation
   bureaucracy 29, 155
   independent agencies 89
   of organisations 136, 159
   public administration 85–6
autonomy
   and horizontalisation 224–9
   in organisations 133
   social domains 33–4
   through ICT 3
autopoiesis 225, 250–2
avant-garde 6, 154, 160, 170, 178

basic registries 66
beauty, administrative aesthetics 237
Big Brother 158, 194–5, 201, 217
BIOS-3 memorandum 127
boundaries, postmodern organisations 184, 197
bourgeois culture 139

bourgeois liberals, postmodern 164–9
brokerage 124
bureaucracy
   amendments to 23–8, 155
   differentiating 28–30, 34, 155–6
   informatisation 56–7, 64–9, 158
   politics 110–15
   postmodern politics 262
   public administration 74
   reversing 30–5, 156
business methods 11, 26–7, 155
business process redesign 135

calculation 41, 71–2, 138–9
capacity, ICT 2–3, 38–9
capitalism 129–31, 139–41, 177
Christian Democracy 15, 122–3, 125–6, 165
church 137
citizen profiles 88
citizenship 105–6
Civil Service, reorganisation 25t, 98
class, postmodernisation 179
classical governance 13–14
co-production, public administration 204–5
coalition-formation 14–15, 118
commodification 140, 145, 147, 170
commodity reification 152
communication
   informatisation 42, 73
   infrastructures 257–8
   postmodern politics 263–4
   see also information and communication technology
communicative instruments 26
communitarism 102–7, 212, 270
community, modern society 176
complexity, informatisation 62
computer conferencing 55
computer coupling 39, 47–9

Index

computer matching 48
computer profiling 48-9, 52
concepts, of modernisation 145-7
conceptual renewal, governance 119
connections
   informatisation 39, 47, 271
   relativism 222-3
   spontaneous 227-8
connectivity, informatisation 39-40, 47-9
constructivism, technology debate 44-6
consumerism 130
consumption 130, 152
contents, network data 47
contingency 223-4, 252-5
contingency theory 148, 264
contract society 141-2
contractual relations 34, 85
control
   classical governance 14
   informatisation 41
   Internet 194
   new governance 12
   public administration 70, 72, 79-80, 94
   through ICT 94, 202
control revolution 57, 75
control systems 65
core departments 22-3, 98
core tasks, independent agencies 20
corporatism 15, 118
coupling, computer 39, 47-9
cultural modernism 170
cultural patterns 136-41, 149-51, 176
culture
   informatisation 56-61
   modern society 176
   postmodernisation 178
   postmodernism 170-3
   technological development 157
cyberpunk 195, 201, 217
cyberspace 196, 206-7, 232-3

data coupling 52
databases 66
De Koning Commission 97-8, 99
de-differentiation 179-80, 213
de-hierarchising 31-2, 32-3, 135-6
de-territorialisation
   independent agencies 90
   of organisations 159
   politics 115, 217
   public administration 86-7
   through virtualisation 198, 201
decentralisation
   bureaucracy 28, 155
   decision-making 121, 134
   independent agencies 16
   politics 114
   public administration 11, 71
decentring
   de-territorialisation as form of 198
   subjectivity 191-2, 200-1, 216, 244, 250
decision-making
   by machines 60
   decentralisation of 116, 121, 134
   grammar for 260-1
   infrastructures 257-8
   political primacy 212
   public administration 80
   socialisation 114
   support technology 66
Deetman Reports 3-4, 97-101, 122, 126
democracy 245
democratisation, public administration 82-3
dependency, welfare state 108
depillarisation 136-8, 150
deregulation, public administration 10
design
   informatisation 37
   postmodern public administration 236
determinism, technology debate 43, 44-6
differentiation
   bureaucracy 28-30, 34, 155-6, 203
   cultural modernism 170
   economic transformations 147
   independent agencies 16
   informatisation 57-8
   modernisation 145, 175-7
   organisational change 148
digital ambiguities
   administration and politics 217-18
   informatisation 61-3
   public administration 92-4
   virtual state 268-71
Digital City 69
digital highways 232-3
discipline, informatisation 41, 72

Disclosure 195
disorganisation, postmodernisation 181-3
displacement, of politics 208-9
domination 61-2, 146
Durkheim 144

economic features, modern society 175-6
economic transformations 129-31
  advanced modernisation 147-8
  nuances and discontinuities 151-2
economies of scope 185-6
Effective Ministries 97
electorate 97-8
electronic markets 130
elites 170-1
End of Ideology, The 166
Enlightenment 5, 167, 246
Enschede, welfare index 49
entrepreneurship 57, 122-3
ephemerality 169-73, 199, 220-1
epistemic enslavement 61
executive departments, informatisation 93
expert systems 66

families 136-7
fashion 141, 187
feedback mechanisms, administration 73
flexibility, postmodern organisations 183
flexible specialisation 132-3, 143, 148-9, 153
formalisation, information systems 51, 56, 75
forms and styles
  postmodern politics 260-1
  public administration 235-8
fragmentation
  and contingency 220-4
  politics 208-10, 250-2
  politics, governance and technology 213-18
  postmodernisation 179-80
  public administration 202-4
fraud 79
front-end verification 48
functional compression 197
functional decentralisation 16
functional rationality 246

differentiation through 5
modernisation 154
and substantive rationality 105, 146-7
technology 2
functionalisation 29, 155

GBA *see* Local Authority Basis Administration
glocalisation 197-9, 216
governance *see* classical governance; new governance; self-governance
grand narratives 6, 165-6, 209, 265
GSDs *see* Local Authority Social Services

Habermas 131, 167, 245, 247
hedonistic pluralism 267-8
Heidegger 166
hierarchies *see* de-hierarchising
homogenisation, of culture 140
horizontalisation
  autonomy and 224-9
  of independent agencies 89-90
  of organisations 53-4, 136, 159
  in politics 215, 216
  in public administration 84-5
  through ICT 3
human resource management 133-4
hyperdifferentiation 177-9, 203, 213

ICT *see* information and communication technology
idealism 247-8
identification numbers 67
identity 136, 140-1, 199
ideology 120
image culture 192-3, 201
image-creation 205, 216
images, social development 141-4
immanent tendencies, politics and bureaucracy 110-12
immediacy 70, 135, 199-200
incentives, steering by 10, 26
independence, public administration 81-2
independent agencies 10
  autonomisation 29
  core departments 22-3
  core tasks 20
  forms 16-17

informatisation 88-91
  motives 17-18
  policy and implementation 20-1, 156
  primacy of politics 21-2
  types 18-19
indifference, political 253-5
individualisation 136-8, 150
individuals, liberalism and communitarism 102
inequalities 107, 177, 179
infocracy 76-7, 158, 198
informating 52, 134
information
  classical governance 14
  growth of, in administration 80
  infrastructures 257-8
  post-industrial society 129
  postmodern politics 263-4
information and communication technology (ICT)
  applications 46-50
  culturalising of the economy 152
  increasing capacity of 2-3
  modernisation 148, 269
  organisational change 134-5
  political implications 126-7, 158-9
  postmodernisation 190-1
  public administration and politics 156-7
information management 37
informatisation
  administration *see* public administration
  bureaucratic culture 158
  characteristics 40-2
  cultural meaning 56-61
  developments 38-40
  digital ambiguities 61-3
  organisations 50-6
  outline of 36-8
  politics and bureaucracy 114-16
Informatisation in Public Administration 7, 69
infrastructures 229-35, 257-8
Inland Revenue 68
innovations, administrative 112-13
insecurity 108
institutional fantasy 122
instrument theory 155
instrumental-professional motives,
  independent agencies 17-18
integration, informatisation 39-40, 49
intelligence, postmodern public administration 236-7
interaction, public administration 71
interdependency, policy making 19
internationalisation 130-1, 140
Internet 47, 57, 193-4, 215
ironic reflection, administrative theory 240-2
irony, postmodernism 1, 266-7

justice, political postmodernism 266

kitsch 169-73
knowledge
  bureaucratic 262
  classical governance 14
  intensity 133, 149, 184
  post-industrial society 129
knowledge-based societies 57

language, postmodernism 166-7
law of policy accumulation 253
liberalism 102, 103, 104, 266-7
libertarianism 31-2
lifestyles 139-41, 151, 173
light-mindedness, moral judgements 169
local administration, informatisation 69
Local Authority Basis Administration (GBA) 68-9, 84-5
Local Authority Social Services (GSD) 67
local determinisms 220
localisation 28, 155

man-machine symbiosis 60, 61
market
  metaphor of the 131, 150
  postmodern organisations 183-4
  relations 136
Marx 110-11, 144
mass culture 6, 139-41, 151, 154, 173
meaning
  in organisations 149
  relativism 222
meta-narratives *see* grand narratives
meta-steering 12, 121, 228-9, 256
metaphors
  administrative theory 239-40

Internet 193-4
market 131, 150
middle management 53
ministerial responsibility 97
mission statements 149, 210
mode of information 45, 191
modern society 108, 175-7
modernisation
  advanced 144-51
  bureaucratisation 56-7, 74
  differentiation 57-8, 175-7
  functional rationality 105
  ICT 269
  politico-administrative system 4-5
  public administration 77-8, 269
  time-space compression 58-9
  utilitarianism 104
  see also postmodernisation
modernism 174
monitoring systems 66
moral entrepreneurship 122-3
morality, aesthetics instead of 168-9
multiformity 14-15, 30
multilateral instruments 26

narratives
  public administration 9
  see also grand narratives; small narratives
neo-corporatism 15, 30, 155
Netherlands
  communitarism 103-7
  politico-administrative system 2, 14-15, 64, 64-5, 117-18
networks
  governance though 10
  horizontalisation 215
  informatisation 47
  public administration 66-7, 81-2, 207
new governance, characteristics 11-12
New Public Management 11
Nietzsche 164, 166
normative politics 105, 212-13, 245-6, 248-9

object systems 65
organisations
  archipelago of 63, 156
  changes in 132-6, 148-9
  informatisation 37, 50-6, 62

nuances and discontinuities 152-4
postmodern 181-7
restructuring 31, 221
technological development 157
time-space compression 59, 186-7, 196-7
output, government based on 10, 26

paradoxical modernisation 202-3
parameters, steering by 205-6
participation, new governance 12
partnerships 10, 85
pastiche, as choice 171-2
performance indicators, governance 10, 26
Permanent Secretaries' report 22
person-orientated instruments 26
personal identification 67
philosophy 167
pillarisation 118
planning 9, 13
pluralising 256
pluralism, political postmodernism 267-8
plurality 106
policy making
  co-production 204
  independent agencies 18-19, 20-1
  informatisation 62, 81, 93
  reversal of bureaucracy 32-3
political postmodernism 264-71
political primacy
  in crisis 207-13
  independence 91-2
  independent agencies 21-2
  postmodernisation 35
  rehabilitation of 211-13
  restoring 125-6
political strategies 121-6
political theory 245-9
politico-cultural paradox 104
politico-ideological motives, independent agencies 17
politics
  administration and technology 1-4
  bureaucracy and technocracy 110-17
  crisis and continuity 95-6
  debate and self reflection 96-101
  fragmentation and connections 213-18
  the great silence 126-7

modern society 176, 177
postmodern 259-64
postmodernisation 179
restored values 101-10
technoculture 200-2
virtual state 270-1
post-industrial society 129
postmodernisation 4-6, 161-6
   ambiguities 187-9
   aspects of 178-9
   bourgeois liberals 164-9
   hyperdifferentiation and fragmentation 175-80
   organising and organisations 180-7
   reversal of bureaucracy 34-5
   social development 155-9
   yuppies, ephemera and kitsch 169-73
postmodernism 161--6, 174t
   ambiguity of modernisation 4-6
   as cultural variant 154
   late capitalism 152
   political 264-8
postmodernity 161-6
power, political 116
pretensions, political postmodernism 265-6
principal-agent relationship 10
privatisation 10
production 177, 179
professional organisations 31, 32-3
professionalisation 113-14, 133-4, 143, 149, 153
professionalism 155
progress 59, 246
proximity, of information 70
public administration
   change, experiment and innovation 9-11
   governance
      classical 13-14
      new 11-12
   informatisation
      bureaucratisation and technocratisation 73-8
      change, experiment and innovation 78-83
      characteristics 70-3
      digital ambiguities 92-4, 269, 270
      systems and applications 65-7
      systems and projects 67-9

   transformations 83-8
   politics and technology, connections 1-4
   postmodern
      autonomy and horizontalisation 224-9
      forms and styles 235-8
      fragmentation and connections 213-18
      fragmentation and contingency 220-4
      infrastructure 229-35
      postmodernisation and postmodernism 155, 202-7
      sovereignty; subsidiarity and corporatism 14-15
      technoculture 200-2
      see also bureaucracy; independent agencies
public-private partnerships 10, 85
purple cabinet 125, 212
pyramidal structure
   fragmentation of 2, 31-2, 250-2
   parliamentary-democratic system 100-1

quantity, of information 70

rationalisation
   of bureaucracy 27-8, 34, 155
   classical governance 14
   cultural modernism 170
   cultural patterns 150
   economic transformations 147-8
   modernisation 145-6
reason 167, 246
refinement, organisational 51
reflexive modernisation 145, 179
reflexivity, administrative theory 240-2
regionalisation 28, 29, 155
regulation, informatisation 62
relational databases 66
relativism 167-8, 221-3
reorganisation 11, 24, 25t, 98, 155
reorientation, politico-administrative system 116-17
representations, of reality 192-3
republicanism 105-7, 212
restructuring 31-2, 221
rhizome metaphor 182, 186

rigidity, informatisation 62
romanticism, futile 107–10, 245–6, 270–1

scale, public administration 233–5
Scheltema Commission 97
science 5, 177, 179, 238–40
scope and meaning, public administration 233–5
sector systems 65
seduction, virtualisation 199–200, 202
self-governance 33–4
self-organisation 33–4, 124, 184–5
self-regulation 12, 29
self-steering 29, 34, 124
service economy 129
service ethos, public administration 79–80
sex relations, postmodernisation 179
Simmel 144
simulacra 152, 171
simulation systems 66
small narratives 264–6
Social Democracy 123, 125
social differentiation 138
social relationships, weakening of 108
Social Security Numbers 67
social signification 264
socialisation, decision-making 114
societas completa 15
sociology, modernisation 144
softbots 199
solidarity 138–9
sovereignty 15, 125–6
space *see* time-space compression
space of flows, the 59
specialisation, flexible 132–3, 148–9, 153
staccato culture 143–4
standardisation, information systems 51, 56, 75
state
  modern society 176
  postmodernisation 178–9
steering
  at a distance 9
  by incentives 10
  classical governance 13, 14
  independent agencies 19
  new forms of 205–6

new governance 12
  toolbox, modified 24–5
  see also meta-steering; self-steering
structural differentiation 146
structure
  classical governance 13
  networks 47
structuring, postmodern public administration 235–6
student grants, automation of 67–8
style
  postmodern politics 260
  public administration 235–8
subjectivity
  decentring 200–1, 216, 244, 250
  ICT 59–60, 158, 191–2
subsidiarity 15
syndrome of cultural modernity 170
system-theory 251
systems, informatisation 65–7

targeting 70–1, 88
technocracy
  aesthetics 261–3
  as art 242–3
  politics 113–14
  public administration 75–6
technoculture 191–5
technological debate 42–4, 44–6
technology
  fragmentation and connections 213–18
  infrastructure 232–3
  organisational change 153
  politics and administration 1–4
  see also information and communication technology; informatisation
telematics 47
Theory of Justice, A 102
three-dimensional graphical images 50
time-space compression
  administration and politics 201
  administrative actions 220
  informatisation 58–9
  internationalisation 130–1
  organisations 186–7, 196–7
totalitarian normativity 248–9
Towards Core Departments 98
tracking 67, 135
traditions 172

transaction systems 66
transcendence 256
transparency, informatisation 42, 52, 72
transport, of data 47
trivialisation, of politics 210
trust, in autonomy 226
truth, political postmodernism 266
Tuning the system 97

Unchained Burgomaster, The 98, 99
uniformity, technological development 62
unintended consequences, politics 252-3
unit-sovereignty 15, 125-6
universalism 264-5
urbanisation 136
utilitarianism 104

value relativism 139
values and norms, informatisation 75
variety 30, 62-3
virtual reality 40, 49-50

virtual state, digital ambiguities 268-71
virtualisation
    independent agencies 90-1
    of organisations 54-6, 159
    politics 115, 209-10
    postmodernisation 195-202
    public administration 87-8, 207
vocabularies
    postmodernism 166
    relativism 222
voluntarism, technology debate 43-4

Wahlverwandtschaft 2, 75, 270
Weber 73, 110, 144, 175, 198
welfare index, Enschede 49
welfare state 138
wellsprings of meaning 105
Wiegel Commission 22, 24, 98, 99, 100, 126
work, postmodernisation 179

yuppies 169-73